D0442919

Great
Mysteries
of
Aviation

By the same author

VIMY RIDGE
KING HENRY VIII'S MARY ROSE
THE QUEEN'S CORSAIR

Great Mysteries of Aviation

Alexander McKee

STEIN AND DAY/*Publishers*/New York

First published in the United States of America in 1982

Printed in the United States of America

STEIN AND DAY / *Publishers*
Scarborough House
Briarcliff Manor, N.Y. 10510

Library of Congress Cataloging in Publication Data

McKee, Alexander, 1918-
Great mysteries of aviation.

Previously published in London as: Into the blue.
Bibliography: p.
1. Aeronautics — Miscellanea. 2. Curiosities and wonders. I. Title.
TL553.M44 1981 629.13'09 81-40499
ISBN 0-8128-2840-2 AACR2

CONTENTS

	Acknowledgments	7
	Foreword	9
1	LOST IN MYTH AND LEGEND Icaros: 1700 BC	15
2	THE ICE FLIGHT OF THE 'EAGLE' Andrée's Polar Balloon: 1897	22
3	THE MONTROSE GHOST The Irish Apparition: 1913–1916	48
4	WHO KILLED THE ACES? Immelmann, 1916; Ball, Guynemer, 1917; Richthofen, Luke, 1918	57
5	THE HIGH SPEED FLYER Sam Kinkead: 1928	102
6	VANISHED The Lost Heroes and Heroines: 1910–1941	120
7	THE MAN WHO WAS NEVER MISSING Helmut Wick, J.C. Dundas and ?: August–November 1940	153
8	THE GHOSTS OF WIGHT 1940–1941	173
9	NO CHRISTMAS IN CONWAY STREET Portsmouth: 23/24 December 1940	189
10	DEATH OF A PRIME MINISTER General Sikorski: 1943	213
11	ABANDONED Flights without a Pilot: 1915–1945	223

12 INTO THIN AIR 236
Joe Kennedy: 1944

13 BACK FROM THE BED OF THE SEA 263
Zuyder Zee: 1917–1950

14 THE LOST AIRLINERS 277
The two Tudors and
The two Comets: 1948–1954

Sources 291

ACKNOWLEDGMENTS

I am indebted to the following for permission to quote from previously published works:

Viking Penguin, New York: *The Andrée Diaries* (copyright 1930, renewed 1958 by Viking Press).
Peter Davies, London: *Sagittarius Rising* by Cecil Lewis.
Routledge & Kegan Paul, London: *Flying Dutchman* by Anthony Fokker.
Verlag Ullstein, Berlin: *Ace of the Iron Cross* by Ernst Udet.
Air Vice-Marshal Sandy Johnstone: *Enemy in the Sky*.
Associated Newspapers Group, London: *Daily Mail* article by Noel Monks.
Blackwood's Magazine, Edinburgh: *Operation Elba Isle* by Gerald Forsberg.

Transcripts of Crown-copyright records in the Public Record Office, London, appear by permission of the Controller of HM Stationery Office. These records are Adm 199/11 Doc 483 and HO 203/ Intelligence Summary 956.

I must express my gratitude for permission to use the portrait of F/Lt Kinkead to the Commandant of the RAF College, Cranwell, and the R.J. Mitchell Hall, Southampton, and to W/Cdr David Bennett, AMRAeS, RAF (Rtd) for the trouble he took in arranging this and other matters.

Also to the many individuals, listed among the sources of information under particular chapters, who drew on their memories and personal papers to help me establish many facts in my narrative.

For help in contacting local witnesses I should like to thank the editors of 'The News', Portsmouth; 'The Isle of Wight County Press', Newport, 'The Weekly Post', Ryde.

FOREWORD

Mankind's conquest of the air almost at once created a new range of mysterious disappearances. Men and women took off in balloons, aeroplanes or airships, and never returned, apparently swallowed up in the vastness of the sky. There was no knowing their last resting places. The sea could have taken some of them, or they could lie in the still depths of deep lakes, on remote mountainsides, hidden in dark forests, or half-buried in the hot sands of the deserts.

Now and then came stranger stories still, to rival those about the abandoned ship *Marie Celeste*, tales of empty aircraft and airships which continued to fly and even made good landings; there was one case of an aeroplane flown by a dead man. In due course there were ghost stories, too. Not phantom aeroplanes as yet, but the apparitions of airmen were said to have been seen. The very first such story dates from early in the century, soon after man learned to fly, and is well attested.

Adding to the mystery was the fact that the air was a comparatively unknown medium. The first aircraft to float or ride on it were subject to forces, either in their design or in the environment itself, which could not then be precisely calculated. Unsuspected but very real dangers lay unseen inside even the toweringly beautiful white clouds of summer. Aircraft really could be swallowed up and destroyed without apparent cause. Yet even in the age of radar and jets things could still go badly wrong, and not always for easily understandable reasons. There were disappearances not explained to this day.

When I was approached to collect a number of such events together to make a complete book, the main difficulty seemed to be in finding a thread to link the episodes. I decided to select good stories which were known to me but not to most people. This meant a bias towards personal knowledge and even personal involvement, but as the mysteries I was familiar with covered the progress of aviation from the earliest days to the recent past, an historical sequence must result. Surprisingly, some of the stories did link directly.

Family involvement with aviation began early. My grandfather first went up in an aeroplane and took photographs from it in 1911, when he was stationed at Farnborough, the home of British aviation in the pioneer days (the first manned aeroplane flew in 1903 in America). I first flew as a passenger in 1931 from a tiny airfield at Bembridge in the Isle of Wight. I started taking flying lessons at Portsmouth in 1933, and was still just a fifteen-year-old schoolboy when I went solo in a Gipsy Moth in 1934. Poor sight in one eye kept me out of the RAF, but I flew gliders and sailplanes in Germany after the war and in 1951 was allowed to pilot a Meteor jet-fighter as part of a radio programme I was producing.

During that long period certain stories stayed in my mind as unsolved mysteries. The earliest was the legend of the 'Montrose Ghost', Lieutenant Desmond Arthur of the RFC. My mother knew Desmond Arthur, who had learned to fly in 1912, before he was killed in a BE2 and came back, it was said, to haunt the old mess at Montrose. I heard the story of his death and his haunting long before it was polished up for publication as fiction, and it had certainly been common knowledge in the Royal Flying Corps by 1917, but how true was it? That is, the original story which I had been told while I was still very young, not the stranger version which came half a century afterwards.

There were other mysteries from the same period. It seemed strange how many of the leading 'aces' on both sides had died in controversy with no one really sure exactly how they had met their deaths, or who had shot them down, or indeed if anyone had shot them down at all. And this in spite of the fact that the writers who discussed the matter had often either been flying wing to wing with the victim a few minutes before, or been shooting at him. In the early 1930s I devoured the British and American magazines which were devoted to the theme of aviation in the First World War and were written and edited mainly by ex-airmen. The British magazines told that Germany's leading ace, Baron von Richthofen, had been killed by a Canadian in a Camel, but one American magazine had another story and was able to quote Australian witnesses who had quite a different tale to tell. Which was the true one? Then there were the disappearances. Captain Albert Ball flew into a thundercloud during a dogfight, and everything after was rumour and conjecture. The Frenchman Georges Guynemer vanished without trace: no crash site was ever found, no grave discovered. And the American balloon-buster Frank Luke vanished likewise (although his fate was later established beyond much doubt).

If the greatest aces and leaders could vanish without trace, how many ordinary men could disappear into the unknown? What were the factors at work to make these things possible? It was only after another world war and some personal experience that I understood.

There were some strange disappearances too of the celebrities of civil aviation in the 1920s and 1930s. The lure which led these men and women to their unknown deaths was the pioneer crossing of great oceans – the grey Atlantic, the frozen Arctic, the formidable Pacific. Frenchmen like Nungesser and Coli, Americans like Amelia Earhart, Russians like Levanesvsky, Australians like Kingsford-Smith vanished into the sky and were never seen again, although sometimes strange rumours circulated. Amy Johnson, who survived many long-distance flights only to disappear over the Thames Estuary during a short ferry trip, was the subject of baseless conjecture and scandal.

The first mystery which I witnessed for myself was the death of Sam Kinkead in the Supermarine S5 in 1928. I actually saw that blue-and-white racing seaplane as Kinkead dived it for an attempt on the world speed record, held by the Italians. A minute or so later, he was dead. A few days afterwards I heard strange stories of what had happened to the pilot. My father, who was a surgeon in the Royal Navy, had discussed the case with the RAF doctors who had been professionally involved. That tale was gruesome and tinged with doubt. Many witnesses thought the whole story had not been told, in spite of testimony at the Inquest. I was to discover that they were right.

This was a local story, because few people outside those involved realised that there was a mystery at all, but it was not parochial. In 1929 *Flight* magazine wrote that the contests between these racing seaplanes were 'the greatest air spectacles in history'. By then the S5 had led to the S6, and not many years later to the

Spitfire. And that in turn led to air spectacles far greater and immensely more costly, staged in the same skies over Portsmouth, Southampton and the Isle of Wight, as the Luftwaffe tried to break the RAF and destroy among other targets the Supermarine works at Southampton.

The dive-bomber units were commanded by Wolfram von Richthofen (who had been aloft on the day the Red Baron was killed) and a new Richthofen *Jagdgeschwader* provided many of the escorts to the bombers which swept over my head in 1940. The new Richthofen Circus, too, lost its leader Major Wick in mysterious circumstances south of the Isle of Wight in a dogfight which I witnessed. Before he died, Wick had shot down a Hurricane on the hill behind where I lived, and I had actually photographed him doing it. The story of that aircraft and its exhumation thirty-nine years afterwards helped explain odd facts about the burial of the Red Baron and the disappearance of Guynemer. There were likely to be prosaic, human explanations for those mysteries.

The Wick mystery was famous in Germany at the time; less so in England where his name was unknown, unlike his First World War predecessor. But another mystery which took place near me was entirely local. To this day, people in Portsmouth speculate as to what it was that at Christmas 1940 caused an enormous explosion near the naval dockyard which destroyed in an instant nineteen streets of houses. I saw the crater myself and heard all the rumours at the time, and was then no wiser. Yet surely this was a mystery capable of solution?

Some stories were so intriguing that I collected cuttings, although I had no personal connection. One such was the awful mid-air explosion in which Joe Kennedy Jr died over England in 1944. The background to that mystery, as eventually revealed, was even more incredible than the tremendous detonation which had been so promptly and efficiently hushed up at the time that even the Kennedy family were not told the truth.

Then there were tales of abandoned aircraft – or of machines with only dead men in them – which had flown themselves. I had read about one particularly chilling incident from the Great War and I was to see two cases, one of them involving a crewless American four-motor bomber which put on an astounding aerobatic show for twenty minutes over the heads of British, Canadian and German troops on the Normandy beachhead. I never did find out the story behind that one, but was more successful with other strange occurrences of the same kind.

When the Zuyder Zee was drained, polder by polder, after the Second World War, it seemed as if the sea was being rolled back to expose the shipwrecks of centuries. The first big surprise was that at first sight there were hardly any wrecks – they were buried – and the second was the number of aircraft from the two world wars which were found, almost every one of which posed a mystery of its own: what was it? who flew it? who died in it? This story is not finished yet. With so many mysteries to choose from I selected aircraft wrecks from two categories: the first, where the investigating team on their own initiative set out to solve the problem; the second, where relatives of dead men asked the Dutch air force to find a particular aircraft and the man inside it.

One very old and little-known puzzle came to my notice when I was researching a few years ago into the loss of the airship *Italia* over the Arctic pack-ice in 1928. The side story was that of three pioneer Swedish aeronauts who had tried to reach the North Pole by balloon in 1897 and vanished from human ken for more than thirty years. Even when their last camp had been found, the mystery of their deaths remained. There were the usual facile explanations, one very ingenious theory, and eventually, long after, the medical truth.

Eventually, it seemed possible to compile a book of mysteries which would span the entire history of aviation from the era of the balloon and the airship to the first jet airliners and from the first stick-and-string warplanes to the weapons of the future, the flying bomb and the V2 rocket. Each story would then slot into a recognisable background and certain themes would repeat themselves in a later period. This seemed to me more satisfying than a haphazard collection of stories, however good each one might be.

There were two further steps to take. The first was to eliminate from my list stories which had already been fully told by other authors elsewhere. Reluctantly, I shed the tale of the financier who had fallen out of an aeroplane and the disappearances of Leslie Howard and Glenn Miller. Secondly, to make sure I did not overlook some classic mystery story I asked around in the aviation world for opinions as to which stories could *not* be omitted. Hot favourite was the Montrose Ghost (which had been top of my own list to start with). Another was the strange disappearances of the two Avro Tudor airliners near Bermuda after the Second World War, which preceded the loss of the two Comet jets: I had this in mind anyway because of the interest of the underwater detective side of that story. Yet another suggestion was the unsolved, still-mysterious accident to General Sikorski's Liberator at Gibraltar.

Basically, these were stories of men and aircraft which took off and never came back; many to vanish forever into the unknown, others to be discovered, and sometimes the mystery solved many years later.

The first of all such stories is some 3500 years old, and as a little conceit I have included it. I wondered long about it when during 1945 and 1946 I was flying gliders and sailplanes in Germany, and two visits to Knossos lately decided me. So I have retold what may be the very first recorded accident in aviation, the death of Icaros during a flight over the Aegean from Crete.

Alexander McKee
Hayling Island, Hampshire

Great
Mysteries
of
Aviation

Chapter 1

LOST IN MYTH AND LEGEND
Icaros: 1700 BC

To me, the prime mystery of aviation has always been, why did man wait until modern times before he learned to fly? Why did he not build flying machines thousands of years ago? I do not mean elaborate constructions involving a number of advanced technologies such as a supersonic airliner or a space shuttle. Nor am I thinking of aircraft which might be practical means of transport. I am thinking in terms of mere curiosity or pure fun, the urge to climb Everest on the one hand, the current fad for wind-surfing on the other, or for that matter, hang-gliding.

Home-made flying machines can be constructed in one of two forms – either lighter-than-air or heavier-than-air. Paper and light debris whirling up into the sky above a bonfire gives the principle of the one, the effortless, rigid-winged soaring flight of many birds illustrates the other method. The basic ideas have always been plain to see, yet apparently no one saw them for say ten thousand years, although many brilliant and highly creative civilisations rose and fell during that time.

The basic materials and the means of working them accurately have also been available for an immense period of time. To construct a hot-air balloon involves some sort of covering – anything from thick paper to extra-thin metal will do; ropes from which to suspend a basket below it; wickerwork or whatever you wish to make the basket, provided it is light; and if you want to be self-contained, then instead of heating the air under the balloon on a static fire, devise a reasonably light and fireproof container for a small fire, so that you can take it up with you. Then, as long as your fuel lasts, you fly. Yet it was not until two centuries ago, in 1784 at Lyons in France, that Joseph Montgolfier made the first actual flight in a hot-air balloon.

The first practical gliders came even later, in France and

Germany, after 1850. A glider is simple to construct, just a wooden framework covered by fabric and with controls operated by thin ropes or wires – child's play for people who could build galleys and sailing ships. The design factors are not quite so obvious, because the effect of different wing plans and sections would require prior experiment, but then the plan and cut of a ship's sails are not arrived at by guesswork – and the whole idea is very similar, the shaping of air to impart efficient motion to a vehicle.

Experiments there certainly were, from the Middle Ages onwards, but they are all slightly mysterious because we have no technical detail as to what type of machine they actually built. For instance an Italian inventor, Danti, made a kind of feather-covered glider and in it flew across the piazza of Perugia; but he pranged on a roof and 'brake hys legges'. Still, a flight of some kind he must have achieved. But it was probably gliding rather than soaring.

Gliding is when the machine descends, more or less slowly, from its start-point, usually a hilltop. Soaring is when the machine rises above its start-point, which implies an efficient machine and some form of natural lift. The two most commonly available types of lift provided by nature are a wind striking a hill at right-angles, which must rise up the slope, lifting with it any bird or light flying machine which can stay in it and not be swept into the downdraught at the back of the hill; and thermal lift, which occurs when the hot sun strikes a surface which heats up more quickly than do the surrounding materials – a cornfield surrounded by woods or water is an example. The corn grows hot, heats the air above, and bubbles of hot air break off and start rising, often to form clouds many thousands of feet above. By circling steadily in such a bubble a bird or an efficient type of glider called a 'sailplane' can rise those thousands of feet too.

I know this very well because I have done it; have even circled in the same thermal as a bird of prey, which was rising slightly faster than I was. Either he was a better flier or he had a superior designer to mine. His construction was bone, flesh and feathers, mine was of wood and wire – nothing exotic about it. Sometimes, however, the air over a large area rises in an extraordinary phenomenon called a 'standing wave'. I was once caught in one of these over Germany, while flying an elementary glider which promptly went up like a high-performance sailplane! And all these forces have always been at man's disposal to harness if he chose, if he could discard the concept of flapping wings, and rise up on rigid pinions.

It was this experience with the clumsy, slab-ended school glider, known derisively as the 'Boat' because that was what its tub-like fuselage resembled, and which nevertheless went up like a lift when the air conditions were right, that made me consider anew what has hitherto always been considered a myth, even if the primary myth of aviation – the fatal flight of Daedalus and his son Icarus. Or Daidalos and Icaros, if one prefers the Greek to the Roman, which in this case is nearer to the source, the Minoan world of ancient Crete in 1700 BC. The legend, if that is what it is, has lasted some 3600 years. Surely there must have been some truth to give the initial impetus to such a tale?

Up to AD 1900 the myth was a simple creation for children. Once upon a time there was a tyrant, King Minos of Crete, victor in war over the Athenians. As proof of his victory, he demanded a yearly tribute of Athenian boys and girls to be sacrificed to a monster, half-man, half-bull, called the Minotaur. To house this fabulous creature, he commanded a Greek inventor, Daidalos, to build a vast labyrinth of masonry, a solid maze of stone. Once inside it, the Athenian youths and maidens would not find their way out, but the Minotaur would find them, and they would die. Then, goes the tale, Daidalos made public the plans of the masonry maze he had constructed for the King, and so an angry Minos put him in prison.

But he escaped with his son Icaros by making the first flight ever recorded which has come down to us (so far as I can tell). Daidalos designed and constructed wings for his son and himself, while still confined in the prison (rather like the glider built by RAF prisoners inside the top-security German prison at Colditz in 1945). Then they took off to fly to Sicily. But the youthful Icaros in his delight at their escape flew too high, rose too near the sun, and so the wax holding his wings together melted and he fell to his death in the Aegean Sea. Daidalos alone reached Sicily.

If one postulates, not flapping wings, but a rigid aircraft of some sort; which either lost lift over the sea (where there are no thermals) or alternatively broke up in the air before falling into the Aegean, one would have a myth no longer, but instead an accident fairly representative of pioneer aviation in the late nineteenth and early twentieth centuries AD. Indeed, after I had flown the old 'Boat' primary glider in the 'standing wave', and stayed up with the sailplanes instead of coming down at once, I was severely ticked off by the RAF instructors because, they said, the strong currents inside this phenomenon could break up a primary glider in the air. However, I got away with it. Icaros didn't.

In 1979 I twice went to visit the site where he and his father were supposed to have made their airborne escape from the legendary King Minos. But he was legendary no longer. When in 1900 Sir Arthur Evans began excavating a hill called Kefala near the town of Heraklion on Crete, almost the only sign of the mythical labyrinth of the Minotaur was the eroded stone of a broken column. The digging revealed not merely a vast palace, now called Knossos, but the heart of a hitherto unknown civilisation which Evans called 'Minoan' after the now no longer mythical King Minos.

The Minotaur was still a bit fabulous, but not entirely, because it was clear that bull-baiting games and possibly some kind of religious rite involving bulls had played quite a prominent part in the lives of the Minoans, who were otherwise very peaceable creatures. As for the labyrinth, that was there all right; it was the multi-story palace itself, which was also the region's administrative centre, a complex maze of offices, storerooms, courtyards and light wells. All of which had been buried for more than three thousand years.

So Minos was a real king, there really was a labyrinth and real bulls were worshipped in it. Why therefore should not Daidalos and Icaros be real Greeks, and aviators to boot? The site is certainly ideal for gliding or soaring, even today when much of the surrounding woodland has disappeared and there is much more bare rock than there would have been then. The sea also was further away then than now, for the Minoan port which served Knossos is now partly submerged.

Visitors are advised to go round the palace in the morning, because the afternoon heat at this site is tremendous. I certainly found it so, but these conditions made it almost ideal as a soaring site, taking advantage of the two main types of 'lift'. The wind from the north which the Greeks call the Meltemi, and which often blows through the Aegean during the summer months, rising up the hillsides around the palace, would provide excellent hill lift nearby. The actual rock on which the palace is built, indeed the very buildings themselves, would heat the air more violently than the areas of shady trees and olive groves. There would be plenty of really strong, buffeting thermals to take a sailplane up rapidly to a great height. Was that what happened to poor Icaros?

The whole island of Crete with its three great mountain ranges crowned by Mount Ida, a supposed birthplace of Zeus, could be soared. On some of those terrifying slopes, nearly vertical, one fancies that a sailplane would go up like a rocket, and staying in the updraught, outspeed any conceivable form of ground pursuit. Or any inconceivable form either, for another extraordinary tale from Crete is that of a robot defender of the coasts, a brass giant called Talos. I take him to be the earliest robot in recorded history (or legend, if you like). A mind which can conceive a robot could design a glider. But Talos too might be more than a myth, because the metalwork and clockwork required seems within the capabilities of the people who designed and built the palace of Minos. To the twentieth-century mind, the most impressive parts of that structure are the water supply and the drains, so ingenious that we ourselves might be proud to have made them.

I suspect that most early efforts at flight failed simply because the inventors tried to produce a mechanical bird. Making a clockwork toy-man which will walk is simple, as every modern child knows, but how to lift a real man off the ground by flapping artificial wings is a problem that still defeats us. Gliding and soaring flight would become possible the moment that someone conceived the idea of riding the air currents in a machine with rigid wings. I am prepared to believe that his name was Daidalos and that his son Icaros came to grief. To me, the only real mystery is the location of the crash site.

There are a number of historians, of whom Apollodorus is the best known, and more than one story. The prime tale is that the pair's destination was Sicily, and that only Daidalos got there. It is hard to believe that he could have covered that distance in a sailplane. There is another, local story that the island of Ikaria is so called (after Icaros) because the world's first aviator plunged into the sea there when his wings failed. Ikaria is near Samos, off the coast of Turkey, not even on the route to Sicily. Off-hand, the contradiction might force one to discard both stories, but there is a clue in Ikaria which points to some underlying truth, because of the rival Ikarian legend which connects the island with Dionysos, god of wine, son of Zeus.

Apollodorus gives us the story of three interlocked groups of people. First there is King Minos and his over-passionate wife Pasiphaë, who had a daughter, Ariadne, by him, and the Minotaur by another coupling, this time with a bull. She managed this feat because Diadalos (his name means 'artful' in Greek) made her a dummy cow which she entered when she wanted to be served by a bull. This tells us a little about Minoan civilisation (perhaps it was over-ripe?), and once more introduces a clever mechanical theme – the artificial wings, the metal man, now the dummy cow. Then enter the second group, Theseus and the doomed children of Athens, and Ariadne falling for Theseus and prevailing upon Daidalos for the secret of the labyrinth which he had built for Minos. It was Daidalos' betrayal of the secret which made Minos imprison him, but what of Theseus and Ariadne?

They went north to the island of Naxos, where Theseus for reasons unexplained deserted Ariadne and returned to Athens. But there the girl took up with Dionysos, and they travelled still further north to Lemnos (they would pass Ikaria on the way). There Ariadne bore him four sons: Thoas, Staphylos, Oinopion and Peparethos – the third group. Two of those legendary names now point to the islands of the Sporades, around 1600 BC, when the Minoans dominated the seas and colonised much of the Aegean. Peparethos is the old name of the island now called Scopelos, where there is an anchorage called Staphylos Bay. On a neck of land by the bay, in 1936, a rich shaft grave was discovered. In it were three bodies, many Cretan artefacts and an immense golden sword, so imposing that it must have belonged to a prince or a governor. His name could well have been Staphylos. He and his brother Peparethos are supposed to have colonised the neighbouring island of Ikos (now called Alonessos), and a site there at Kokkinokastro, similar in layout to Staphylos Bay, is pointed out today as being probably the site of the ancient city of Ikos itself. Both sites make very great sense in terms of shipping and the prevailing wind directions in the Aegean, but neither has been anything like fully excavated. In 1978 I explored parts of both these anchorages underwater and found signs at Kokkinokastro of undated submergence and at one point too deep for me what looked tantalisingly like the ground plan of a Minoan house.

Sir Arthur Evans showed around 1900 that King Minos was no legend but a historical character. More lately, we have come to regard two of the legendary sons of his daughter Ariadne as being historical personages also. So why not Diadalos and Icaros?

However, even if I have been giving too free a rein to imagination, if all is fancy and there never was a Greek glider pilot's son called Icaros, who fell from the skies over the Aegean 3600 years ago, yet the youth is real in one sense. He stands for the man who flew too high, dared too much – and died.

As did most of the men and women whose stories I have yet to tell in this book.

Chapter 2

THE ICE FLIGHT OF THE 'EAGLE'

Andrée's Polar Balloon: 1897

The comments in the Austrian newspaper were biting: 'This Mr Andrée, who wishes to go to the North Pole and back by means of an air balloon, is simply a fool or a swindler.'

Indeed, the project did appear fantastic, as Andrée outlined it in 1895. Based on fast speeds obtained over short distances by balloons, he estimated that, from a forward base at Spitsbergen, a balloon could reach the North Pole in ten hours, although an average wind speed of 27 kilometres (16.2 miles) per hour would increase the time to forty-three hours. His intended journey across the Pole to the Bering Strait, a direct distance of 3700 kilometres (2220 miles), would take not more than six days, well within the capacity of his balloon, which he estimated could cover 19,400 kilometres (11,640 miles) in thirty days.

The balloon, which was to be spherical, had no engine. For movement, it depended entirely upon the strength of the wind. For direction, it depended upon whatever wind happened to be blowing, modified slightly by an arrangement of sails combined with ropes intended to drag along the ground. The draglines would ensure that the balloon moved a trifle more slowly than the wind, thus enabling the sails to take effect. Without such a device, the sails would be quite useless. Andrée had satisfied himself, by making a test flight across the Baltic, that the method would more or less overcome the prime drawback of all balloons hitherto, that they could not be steered. The lifting gas was hydrogen, and ballast was carried to counteract both the effect of long-term leakage through the fabric and temporary loss of lift caused by falls in temperature.

Salomon August Andrée, son of a Swedish apothecary, was born in 1845, so he was forty-two at the time of his Polar flight in 1897. A scientist by training, he developed a passion for aeronautics, which he studied in Philadelphia and in France from 1876. His plans therefore were based on twenty years' experience, including many actual balloon ascents. He was now Chief Engineer of the Patent Office and was able to marshal support from prominent scientists and polar explorers as well as from Alfred Nobel, the inventor of dynamite. He was a very stubborn man. When told that there were legal objections to a proposal he had in mind, Andrée answered: 'Well, then the law is wrong and shall be altered.' His reason for not marrying, although well-off, was that a wife would introduce incalculable factors into his life, which he would not be able to control because he would not have the right to do so.

When Nobel had invented dynamite the Swedes, intoxicated with this new-found power over nature, blasted out the rocky landscape of the Stockholm area in an explosive orgy. Andrée went further. He was one of those – and they were many – who believed strongly in the inevitable development of technology and its progressive application to improving not merely the lot of mankind but man himself, through intellectual and scientific training. Workmen would become more eager and intelligent, and everyone would be happier. As he was himself a philistine, Andrée did not desire to inflict art, literature or music upon them, but otherwise he was representative of an outlook common in many European countries, and not entirely extinguished even to this day. Its keynote was optimism. In Andrée were combined two of its basic beliefs, namely that the two final conquests of nature were about to be achieved: the conquest of the air by directed, sustained flight (as opposed to free ballooning or simple gliding), and the attainment of the North and South Poles by one means or another.

Andrée's was by no means a foolish optimism. He struck a balance: 'The thing is so difficult that it is not worth while attempting it,' or: 'The thing is so difficult that I cannot help attempting it.'

He chose two much younger men to go with him on the flight. The first was Nils Strindberg, aged twenty-four, a relative of the author August Strindberg. He had studied physics, chemistry and music, had been a photographer from the age of sixteen, and went to Paris in 1896 to obtain experience of balloon ascents. He was engaged to a girl called Anna Charlier.

The third place was taken by Knut Fraenkel, aged twenty-seven.

A keen sportsman, and the son of an engineer officer, he had been kept out of the army by a nervous complaint and had studied civil engineering instead. Together, the three men made up a balanced team.

On 11 July 1897 a strong southwesterly wind was blowing over the anchorage at Danes Island, Spitsbergen (now called Svalbard). An essential feature of Andrée's plan had been a base here, on the fringes of the Polar ocean, where the balloon would be housed and finally filled with gas. The winds had been unfavourable the previous year, and for most of this summer. Even this present wind would not blow them to the Pole but northeastwards, perhaps to somewhere off the coast of Siberia, unless it became more favourable later. Andrée had made great study of Polar winds, and now he was troubled. The wind was good, but was it good enough? Faced with an immediate decision, he may have realised with a stab of alarm that although his plans seemed sound, too many were theoretical calculations only. There had been no lengthy period of long-distance flights to test either the balloon or the idea. Admittedly, he had flown the Baltic, but the journey ahead was ten times longer and far more severe. He turned to his companions.

'Shall we try or not?'

Fraenkel, the young sportsman and engineer, was unsure and unwilling to commit himself. At length, he agreed they should start.

When Strindberg was asked, he replied: 'I think we ought to attempt it.'

Andrée conveyed his uncertainty to Captain Count Ehrensvärd, who commanded a royal warship supporting the project. He told him: 'Well, we have been considering whether to start or not; my companions insist on starting, and as I have no absolutely valid reason against it I must agree to it, although with some hesitation.'

Strindberg was occupied until the last moments taking photographs of the start. Then Andrée cried: 'Strindberg and Fraenkel, are you ready to get into the car?' The balloon was now held from rising only by three lines attached to the ballasted car beneath it. Calmly, without expression, Andrée called out: 'One! Two! Cut!' Simultaneously, three sailors slashed through the ropes and the balloon rose majestically to shouts and cheering. At the moment of flight, Andrée christened it the *Ornen*, or 'Eagle'.

But it did not fly like one. All estimates of its height place it at less than 300 feet. Then, because it was in the lee of a mountain, the balloon was caught in the downdraught and swept erratically towards the surface of the bay. There were cries of alarm from the sailors who, unlike Andrée's team, had not expected this; they ran to the shore to launch rescue boats. The balloon's car struck the water, then bounced into the air again. There was a shout from a sailor: 'Why! the draglines are lying here on the shore!'

It was true. They had not been broken, but had become unscrewed from the metal connecting points by the curling and uncurling of the lines as they slithered across the rocks and stones. They had been an essential element in Andrée's scheme to make the 'Eagle' semi-dirigible. The drag of these ropes along the ground would slow the balloon enough to allow the sails to fill with wind and, using those sails for steering, Andrée had reckoned that he might be able to fly as much as forty-five degrees off the wind. A fault which was not merely unforeseen but freakish had doomed the expedition from the start.

The balloon, low over the water, dwindled in the distance until it appeared no larger than an egg. It was making a good speed of around 20 mph, and the sailors became agitated again, thinking that the 'Eagle' would dash itself to pieces against a distant hill. Of course, Andrée's colleagues knew that the updraught must carry the balloon clear, as indeed it did. Almost their last view was of the 'Eagle' rising above the hill and vanishing beyond. Briefly, a speck appeared, seen between two hills, speeding away in the distance over the Polar sea. There was nothing more.

Andrée's expedition was a few years too early for practical wireless communications to be possible. The 'Eagle''s message-sending capabilities consisted of thirteen buoys and thirty-six carrier-pigeons. Three buoys were picked up – two in 1899 and one in 1900. Two of them carried messages dated 11 July 1897, the first day of the flight; the other carried no message at all. Only one carrier-pigeon got through, with one message (they were supposed to bear two), and died delivering it.

On 15 July 1897, the skipper of the Norwegian sealer *Alken* was called on deck to see a 'peculiar bird' which, coming from the south pursued by two ivory gulls, had taken refuge on the mast. The skipper climbed the rigging and shot the bird, which fell into the water. He did not bother further with it until the *Alken* heard from another ship later that day about the Andrée balloon and its pigeon-post. He immediately went back and sent out boats to search the water; one of them located the dead pigeon, still floating. It did indeed carry a cylinder with a message inside:

> From Andrée's Polar Expedition to *Aftonbladet*, Stockholm.
> 13 July
> 12.30 midday, Lat. 82° 2′ Long. 15°5′E. good speed to E. 10 south. All well on
> board. This is the third pigeon-post.
>
> Andrée.

This was the first, and for some years the only, direct news from the occupants of the Polar balloon.

Of course there were sightings, and rumours of sightings. The first was from Manitoba, Canada, on 1 July, ten days before the 'Eagle' had flown at all from Spitsbergen. On 17 July a Dutch ship hundreds of miles to the south of Spitsbergen, whose captain knew nothing of Andrée, reported a large object some 150 feet long floating on the water and not clearly seen because of haze. He thought it might be a small vessel drifting upside-down. Some newspapers assumed that it signified the end of Andrée. Others later dismissed the object as more likely a dead whale. However, the object did bear a similarity to a sighting made on 13 July by the bark *Ansgar* and reported in a Danish paper. Five men of the crew certified that, two days east of North Cape, they had passed a balloon at a distance of ten fathoms (sixty feet). 'It was black in colour, and some of the gas had leaked out. It looked about two fathoms [twelve feet] high above the waterline, and was covered with a net.' Even the positions given by the two ships tended to corroborate each other: they might well have seen the same floating object, a few days apart.

The carrier-pigeon message recovered on 15 July contradicted both sightings. The position given in Andrée's writing and dated 13 July was well to the north of Spitsbergen, towards the Pole, not southwards off the coast of Norway. Perhaps it had been a bloated whale after all.

There were other stories of a more tenuous character. An elderly lady 'whose truthfulness is beyond question, says that on the evening of 17 July she was about to go to bed, and had gone to her window to pull down the blind, when she noticed something which looked exactly like a balloon with drag-ropes and a net. It had a gondola with apparently one man standing in it.'

There was a chilling story of the screams heard in the surf. Captain Johan Overli of the sealer *Swan* had been off Dead Man's Island on the coast of Spitsbergen on 22 September, when he and the whole ship's company had heard a single scream above the noise of the surf, and then three more cries in quick succession. The seas were too heavy for them to go ashore and shortly after their own ship was stranded. They were rescued by another ship, the *Malygin*, which had to pass Dead Man's Island on her passage to Norway, and the crew of this ship also heard renewed screaming from the sinisterly-named Dead Man's Isle. The captain refused to investigate, saying that the noises were only the screeching of birds. So they might have been, but on 23 September there was independent corroboration, or so it seemed, for another captain, Olaus Olsen of the *Fiskeren*, had on that day been only a few miles away from where the screaming had been heard, and he reported seeing wreckage which he was unable to approach. At the time, he had not thought of Andrée, but now it seemed that his sighting could have been the remains of the 'Eagle'. A Norwegian search of the area discovered nothing.

On 25 September came a report from a different locality – Ivigtut, in Greenland. The bark *Samia* brought a tale to Philadelphia that the natives had seen a balloon drifting north at 1000 feet some three weeks after Andrée had left Spitsbergen. There was seeming corroboration for this tale also. Captain Bang, the master of a Danish ship, reported that the natives of another Greenland settlement, Angmagssalik, had heard a gunshot out on the ice in late October or early November. The East Greenland Current does carry ice-floes southward throughout the year, and it did not seem impossible that if Andrée and his companions had landed on such a floe, they might signal to the land by firing a gun. On the other hand, pack-ice naturally makes noises easily confused with gunshots.

A quite different type of story began to come out of Imperial Russia. Instead of tales of screams or shots, there were sightings, and not of wreckage only. A newspaper in St Petersburg printed a report jointly signed by an engineer and an academic that on 20 September they had seen a balloon drifting over the town of Yakovlevskaya. It had a curious, electric sheen about it and was very small (about seven inches, unless the Russian measurements have been confused); it was visible for under three minutes, moving rapidly. The direction of flight was towards central Siberia, and this lent credence to a slightly earlier report that a balloon, thought to be Andrée's, had been seen on 14 September over the Yenisei valley in Siberia, drifting northwest. The Russians would have known what to look for, because the Swedish authorities had circulated descriptions and pictures of the missing balloon. However, Andrée had calculated the flight time of his balloon as not less than thirty days. If these stories were true, the balloon had been aloft for more than double that time. It could hardly stay up much longer.

Consequently, one could believe reports which began to come out of the same area of Siberia, the Yenisei, in February 1898. They seemed to be based on the discovery by nomads of wreckage and unknown instruments, and nearby the corpses of three men, one with his skull crushed. This story attracted worldwide attention and the Swedish authorities urged the Russian government to investigate. A painstaking detective inquiry, lasting months, finally established that the initial report had been of a strange discovery by natives of three dead horses. The speculation this caused locally had spurred a hoaxer to turn the find into three men instead of three horses, adding graphic detail.

With Central Siberia as the crash site now discredited, it was possible to believe stories coming out of an area thousands of miles away, the province of British Columbia on the Canadian Pacific coast. True, Indians had told of seeing a balloon drifting to the north early in July, before Andrée and his companions had even left the ground. But there had been a later report, from the first week of August, of a Mrs Sullivan and her daughter telling a Mrs Hobson (who told Mr Hobson, who told Mr Newson, who told a Swedish consul) that she had seen a round grey object in the sky which came nearer, sinking slowly, until she could see that it was a balloon with a basket suspended below it. Then, near Quesnel Lake, it had moved away and begun to ascend.

At roughly the same time, early August 1897, a series of telegrams began to reach Sweden from the small town of Germania in the state of Iowa, USA. The first one read: 'Andrée moving southwestward near longitudinally west towards Edam Land. Ole Bracke.'

So Andrée was still alive! But Edam Land, it turned out, was in Greenland. And Mr Bracke was living in Iowa, USA. To further inquiries, Bracke replied: 'Andrée makes for safety, seeking whalers.' When asked how he knew this, was he sure, Bracke cabled: 'Yes. Consider my former cables indicative Andrée's situation.' When asked, was Andrée safe, he replied: 'Assistance wanted Andrée. Search coast Edam Land.'

Instead, reporters descended on Mr Bracke, who lived in a small house out on the prairie. He proved to be the author of a poem about the Andrée expedition which had been published in the Minneapolis *Tribune*. He had seen Andrée three times, he hinted. In visions. These experiences were very real to him, although he seemed reticent over detail; but all three men were still alive, with their equipment, he said. And the reason for his telegrams was to convince wealthy men to set up a search expedition to save them.

A number of expeditions did set out in 1898. Stadling's Swedish party searched the Yenisei Valley; Wellman's USA party investigated the Franz Josef Islands north of Russia; Nathorst's Swedish party visited Spitsbergen and White Island, and in 1899 explored North-east Greenland. The expeditions did carry out some scientific work, but there was no sign of Andrée or his balloon. The blank drawn at the Franz Josef Islands was particularly disappointing, because there was a food cache there, left by a previous expedition, and if Andrée's balloon had come down it was felt that he and his companions would try to 'walk out' of the Arctic via Franz Josef Land.

Of Andrée there was no authentic word or sign until 14 May 1899. On that day, in a fjord on the north coast of Iceland, a buoy was found washed up. After 672 days, it was not undamaged. There was a rent in one side and a small copper flagstaff was missing, as was the flag. But the message it contained was readable:

BUOY Number 7

This buoy is thrown out from the Andrée balloon at 10.55 p.m. G.M.T., July 11, 1897, at about 82° N. Lat. and 25° E. Long.

We are floating at a height of about 600 metres.

All well.

Andrée Strindberg Fraenkel

This was an intensely disappointing find. It had been thrown out of the balloon on the evening of the very first day, two days before the release of the carrier-pigeon which had been shot and its message of 13 July recovered.

Even more disappointing was the discovery of another buoy, this time in King Charles Land, Spitsbergen, on 11 September 1899. There was no message with it at all. This was the large so-called 'Polar Buoy' which Andrée had intended to drop over the Pole or at the 'furthest north' attained by the balloon. It could very well have been jettisoned to save weight at some time of crisis for the aeronauts.

Then, on 27 August 1900, a woman who was collecting driftwood on the north coast of Norway came across another buoy, quite intact, which had only just come ashore. When opened, it proved to contain, not the latest news of Andrée, but the oldest. Its message read:

BUOY No. 4, the first that was thrown
11 July 10 o'cl.p.m. G.M.T.

Our journey has hitherto gone well. We are sailing onwards at a height of about 250 metres in a direction at first towards N. 10° East due course, but later towards N.458 East (sic) due course. Four carrier pigeons sent off 5h.40 p.m. Greenw. time. They flew westwards. We are now over the ice which is much divided in all directions. Weather magnificent. Humour excellent.

Andrée Strindberg Fraenkel
Above the clouds since 7.45 G M T

After that, there was no news. And no hope. As Wellman wrote in the February 1899 issue of *Century Magazine*: 'Poor Andrée! poor, brave, dead Andrée!' He would never be found now. The likelihood was that the balloon had been forced down on the drifting pack-ice somewhere between Spitsbergen and the Pole. When the ice drifted into warmer waters, it would have melted and Andrée and his companions would have vanished into the black depths of the Polar Sea.

A World War began, and ended. There was Revolution in Russia, and then in Italy. The story of the Polar balloon was forgotten, for now there were Polar airships, and in 1926 the airship *Norge*, with a part-Italian, part-Norwegian crew, crossed from continent to continent via the North Pole.

In 1928, the airship *Italia* made a series of Arctic flights from a base on Spitsbergen and crashed on the ice coming back from the Pole. Lightened by the loss of its control car, the hull of the airship with six men still aboard rose into the air and drifted away over the pack-ice, never to be seen again. From the wrecked control car emerged nine men, two badly injured. For a time, they were all given up for lost, but eight of them lived long enough to be rescued – by aeroplane or by icebreaker. Three men had attempted to 'walk out' over the pack-ice to Spitsbergen or one of its offshore islands which they could see, but at that time of the year the ice was melting and there was impassable open water between them and solid ground. It was assumed that the hull of the *Italia* and the six men in it had gone the way of Andrée's party and the balloon *Ornen* in 1897.

In 1930 a joint scientific-cum-sealing expedition set out in the Norwegian sealer *Bratvaag* to explore Franz Josef Land. On the way they had to pass a mysterious and usually inaccessible island northeast of Spitsbergen now called White Island because of its virtually permanent capping of ice and snow. On that summer day of 1930, there was a mirror-calm sea and on some of the low-lying parts of the island next to the water, the snow had melted and the rock was bare. There was an intense and deathly quiet about the place, made more ominous by the occasional reverberations from the north, where a melting glacier was loosing hundreds of tons of ice into the sea. It was the sighting of walrus on 6 August which brought a decision to land two parties on White Island – one composed of scientists, the other of sealers. These were largely unexplored waters, so the sealer anchored way out in nine fathoms and the men went ashore in boats.

Boat No. 2 under Sevrin Skjelten chased the walrus south, killed several, and towed the carcasses to shore for skinning. The crew were all local men, from Aalesund on Spitsbergen, and they found it hot work in the Arctic sun. Two of the lads, seventeen-year-old Olav Salen and twenty-four-year-old Karl Tusvik, walked along the strand in search of a drink of water. They found a stream in that icy wilderness and, wading across to the far side, saw an astonishingly incongruous object lying on newly-bared ground – a pot-lid of aluminium. Looking round the half-melted snow and ice inland they noticed something dark showing against the white. It was the upper half of a light canvas boat, of the sort Arctic expeditions carried with them. The lower part they could not examine properly, because it was encased in ice, but part of the equipment had been a brass boathook. It was barely two years since the hull of the airship *Italia* had disappeared some hundreds of miles to the north of this place, and the boat could have been part of her emergency equipment. They went back to the beach and told the *Bratvaag's* skipper, Peder Eliassen, in boat No. 1, what they had found.

He accompanied them back to their find and noticed, about ten metres still further inland, lying on bare rock, the frozen corpse of a man, the legs still half-covered by snow, inside a pair of Lapp boots. The upper part of the body had been much disturbed. Bones stuck out of the clothing and the head was completely missing. Probably bears had been at it. Eliassen, looking around for some means of identification, found it in the boat, where much of the equipment was marked with the name and date of the expedition concerned. Among the jumble was a book, damp and heavy with water, most of the pages stuck together; but openable in one or two places to reveal neat, highly legible handwriting. This was the evidence which he took back to his ship to show the sceptical scientists.

The scientific chief of the *Bratvaag* was Dr Gunnar Horn, a Norwegian geologist. Eliassen 'calmly and quietly told us that they had made a great find', recalled Dr Horn. Then he showed them the book, opening the first page. Part of the book's title was still readable there: 'The Sledge Journey 1897'.

It was Andrée! However, the writer of the book must have been Strindberg, they thought, because he had been the scientific member of Andrée's party and the entries were in effect the navigational log and astronomic calculations made by a party of men undergoing what had turned out to be a death-march across the terribly jumbled and hazardous terrain of the pack-ice. Yet all the entries were so orderly and neat that they might have been made at leisure, by a man sitting at a desk in a warm room.

In their turn, the scientists went up to Andrée's camp. It had been pitched thirty-three years before in the lee of a rocky hill, and so the snow had built up over it, dispersing now only in this unusually warm summer of 1930. There had been other visitors to White Island since 1897, including Nathorst's expedition actually looking for Andrée, and others in 1928 searching for General Nobile and any survivors from the *Italia*, but probably there had

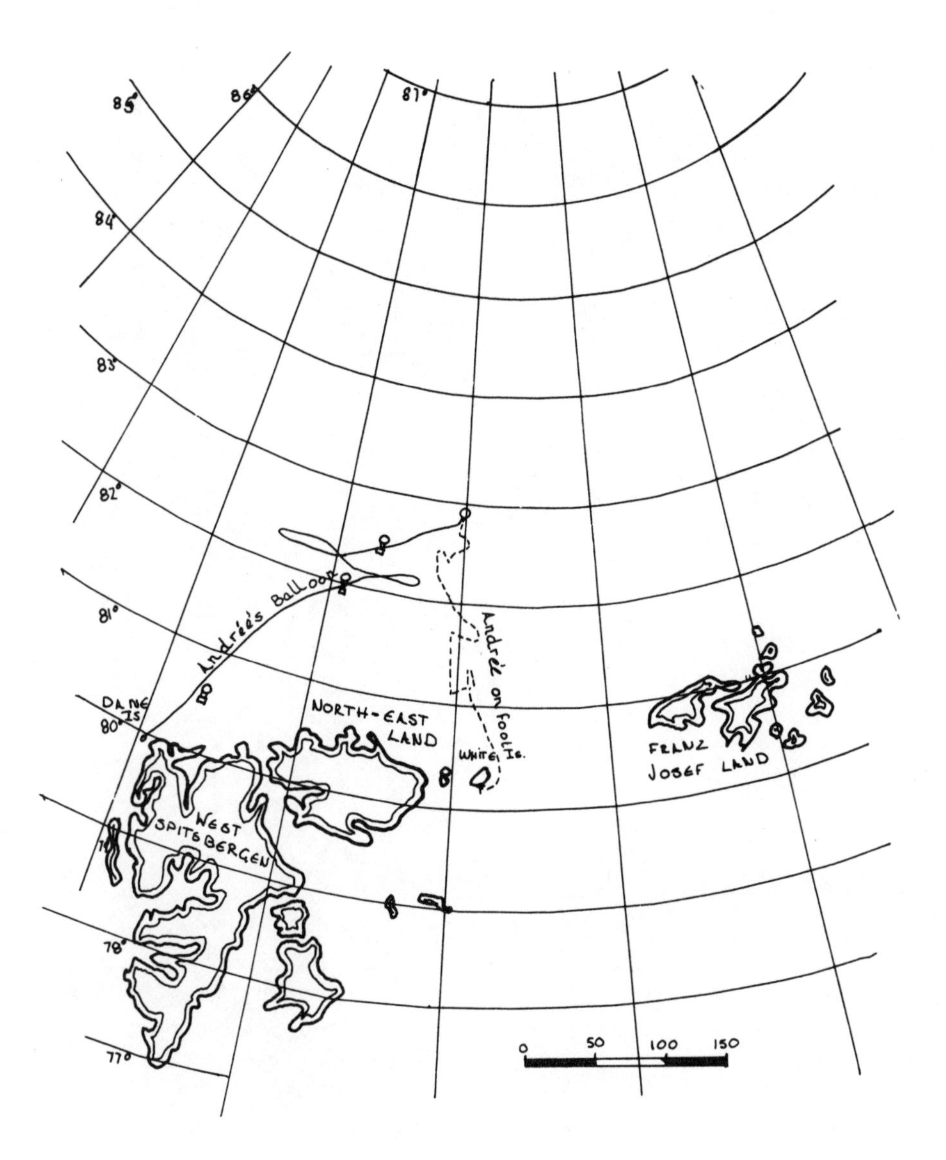

Andrée's Polar Expedition, 1897

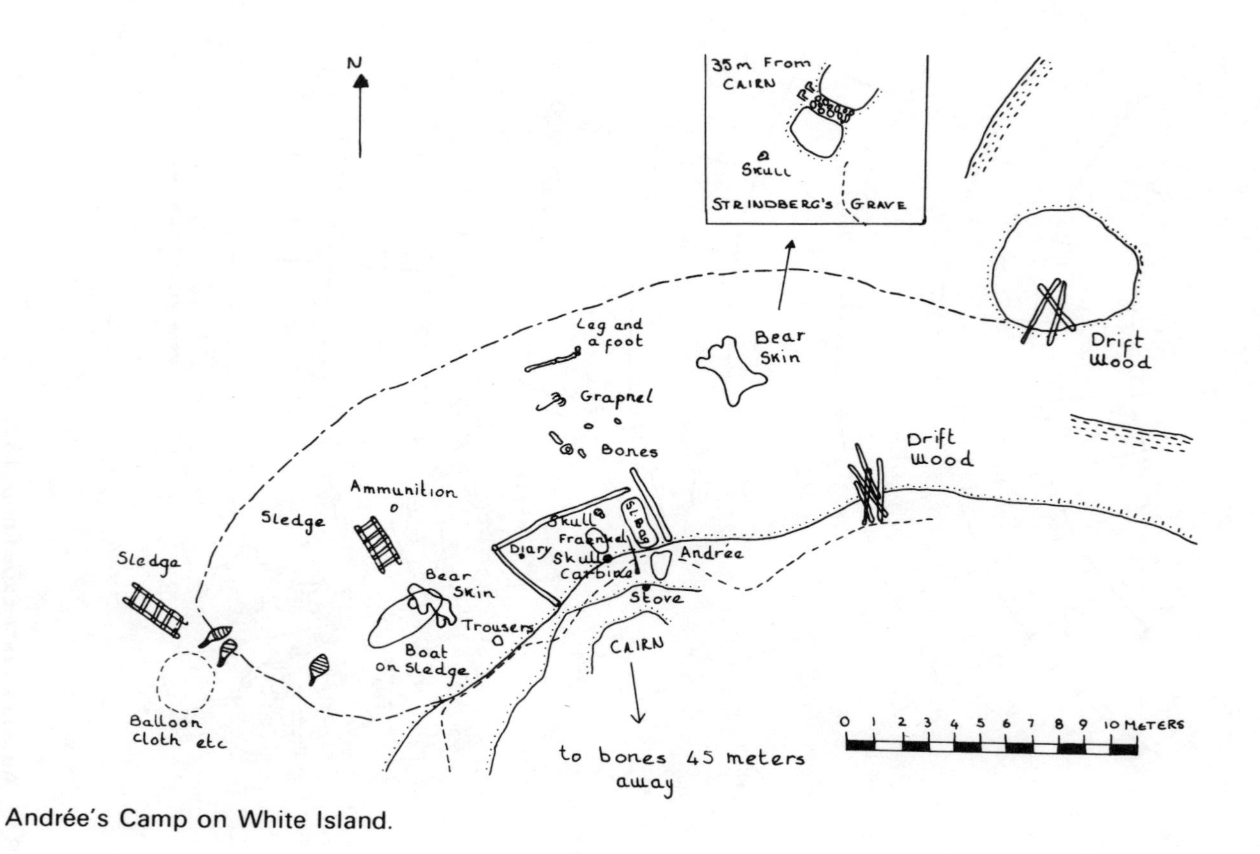

Andrée's Camp on White Island.

then been nothing for them to see. Now there was a litter of material, in and around the boat; also an empty sledge and the bones of a Polar bear. Two shotguns and a rifle half-buried in the snow indicated that the bear had been killed for food by Andrée and his companions before they had died and themselves become food for bears. Still, although the upper part of the frozen corpse was missing, its jacket remained, with a diary in one pocket. The scientists, bending down, carefully opened the jacket and inside the collar saw, sewn into it, the monogram 'A'. So this sorry thing was Andrée! But where were Strindberg and Fraenkel?

On top of some clothing was a Primus stove, with paraffin in it and a valve in working order. There was a cup, cooking utensils and an axe nearby. Equipment of all kinds from a harpoon to a hammer and a Swedish flag lay scattered to some distance, and sixty metres away to the east lay a pelvis, which they thought to be Andrée's. Then Sevrin Skjelten, the harpoonist of No. 2 boat, searching to the north thirty metres away, saw a bleached cranium lying among stones. Looking more closely, Skjelten discovered a second corpse. Unlike Andrée, who seemed to have died where he lay, this man had been buried in typical Arctic fashion where digging is impossible. The body had been placed between rocks and then covered over with stones, to keep off small animals. But bears had been at this one, too. The feet, in Lapp boots like Andrée's, stuck out from under the bottom of the pile of stones and up near the top, above the stones, lay a shoulderblade. When the stones were removed, it was seen that this corpse was headless also, and frozen solidly to the ground. From initials still visible on the clothing, the scientists assumed that this was Nils Strindberg, the scientist and photographer, who had been engaged to Anna Charlier. The very engagement ring lay there on the rocks.

They debated a long time what to do. Should they leave the poor remains to the endless, icy silence of the Arctic? Or should they collect them up and return the bodies to Swedish soil?

At length, the decision was for the latter course and they began by digging out the boat and the sledge on which it proved to be lying. The contents of the canvas craft included books, but to avoid damage everything was left as it was, except that the lashings which held the boat to the sledge were cut. Then the two bodies were put into tarpaulins, and with the boat and the sledge carried the 200 metres down to the water, and so out to the sealer.

In spite of the strange evidence of the Primus stove, which they had tested and found in working order, Dr Horn was to write of the camp on White Island, 'here they fell victims to exhaustion and cold.' There was no Andrée mystery any longer, they thought. Everything was now known.

The man who brought this news to Norway was Gustav Jensen, skipper of the homebound sealer *Terningen*, to whom the outbound *Bratvaag* had passed the message. These little craft did not have wireless transmitters, only receivers. A mad scramble by the press to get to White Island was won by the experienced journalist Knut Stubbendorf, representing two Scandinavian newspapers and an American syndicate, who had earlier covered the Nobile tragedy. In trying to get a ship to take him, Stubbendorf encountered skipper Jensen of the *Terningen*. 'He calmly assured me,' wrote Stubbendorf, that 'he had recognised Andrée, where he lay frozen in clearest ice on board the *Bratvaag*.' In fact, the broken remains of the balloonist had no head attached, as we know. Another Tromsö skipper, Theodor Grödahl of the *Hanseat*, must have been chagrined at the news. On 9 July, less than a month before the discovery by the *Bratvaag*, he himself with two men had searched that part of White Island for the wreckage of Nobile's *Italia* and had noted a sheet-iron box and a wooden peg which must have been Andrée's but which he had ignored because they clearly were not parts of an airship.

Even if he got to White Island, Stubbendorf did not expect to find much, because already a Scientific Commission (hostile to journalists) had been set up by the Swedish and Norwegian governments at Tromsö, and they seemed to know everything, including the fact that the body of Fraenkel would be discovered in the bottom of the canvas boat, unless of course he had died out on the ice before the party had ever reached the shore. Stubbendorf would have been happy just to get a simple interview with the discoverer, Dr Horn, but in this he failed. He did however manage to charter the sealer *Isbjörn*, and although last in a race to get to White Island, in fact he led not merely the first but the only press expedition actually to land. He and six others went ashore on 5 September and made for a cairn which Dr Horn had erected to mark Andrée's camp. He was in light-hearted mood, a journalist freed from a desk and with no responsibilities except how to write a piece about the surroundings where Andrée had died, and to ponder why and how he had died.

A sheet of disintegrating ice covered part of the camp which, it dawned on Stubbendorf, must have been invisible to the party from the *Bratvaag*, for there lay yet another sledge surrounded by clothes and other human debris. Three men, three sledges, that was the arithmetic; and there was the third. Now, Stubbendorf knew that he was in the position of an archaeological excavator: he must remove nothing without photographing and recording beforehand, and then only after careful thought. By so doing, he made sense of the camp. He was able to distinguish a tent or living area, and a separate parking place for sledges and gear. The living area had been much disturbed by bears, but almost at once he discovered a backbone, a pelvis and one thigh-bone; and then, nearby, a shoulderblade. That was on the first day. On the second day, at the same place, but on a lower level, a more complete skeleton consisting at first of a whole bone, a thighbone, a kneebone and a foot; a little later, the bone from an upper arm, with rags of striped shirt around it; finally, seen dimly in the ice, a human skull.

Almost the whole of this second day was spent in carefully clearing around these remains, to reveal the upper body and head of a man lying on his left side, left hand beneath his head, frozen fast to the ground. Stubbendorf had a timber coffin made for these remains, which were removed intact.

He only had half of one more day, for the weather blew up most dangerously for the waiting *Isbjörn*. On his last scout round, Stubbendorf found a pair of perfectly-preserved snowshoes which had been concealed by the ice only the day before. It was a nice discovery, but as nothing to the finds made on the very first day. They comprised what were probably the bulk of the expedition's documents – an almanac and three memorandum books kept by Fraenkel, a logbook kept by Strindberg, and two tin boxes full of films – many of them exposed. The whole story of the Andrée expedition, fully illustrated, was potentially what Stubbendorf had in his cabin as the *Isbjörn* steamed away. Few journalists can have experienced such a scoop. What had happened to the three men, after they had sailed away in the *Ornen* on 11 July 1897? It must all be there, in those sodden lumps of leather and paper. The story of the flight, the story of the trek across the ice. If only Stubbendorf could prise apart those pages without tearing them.

They were frozen, so he let them thaw in the cold, open air. Then he had them brought to his cabin, where they smelt vilely, but at a certain stage of evaporation the pages reached a point where they were neither too soft nor too hard, they neither dissolved nor tore – and then they could be separated by carefully inserting the point of a knife. But once separated, they would fuse again unless each page was detached and placed on a piece of coarse paper to complete the drying process. With the first document – Strindberg's logbook – Stubbendorf worked for twenty-four hours without sleep. And he was rewarded by success. Strindberg's photographs, of course, he dared not touch. But he needed time in which to deal with all of the documents, and planned to take the ship into a quiet fjord. He had barely arrived there when the Scientific Commission at Tromsö ordered him to hand over all his finds to them at once and sent two warships to enforce the order. The barely-concealed fury of the academics is plain in their official report, in which, from sheer force of circumstance, they were forced to include a contribution by Stubbendorf because he had carried out valuable work which could not be ignored.

They had Andrée's diary, which they had preserved in much the same way as Stubbendorf had used with Strindberg's – only they had blotting paper to insert between the pages instead of ordinary paper. They were also embarrassed by a memorial service which they had organised for Andrée and Strindberg in Tromsö Cathedral – leaving out Fraenkel. That had to go ahead. Although they had discovered that some of the soft parts of the bodies had survived inside the clothing, the Commission were now aware that they really had very little of Andrée. The upper part was almost entirely missing. When the Isjbörn's finds were delivered to them, it was decided that these consisted of the upper half of Andrée, being entirely fleshless bones, and also the upper half of Knut Fraenkel, with original clothing, one item bearing his initials. These remains were put into zinc caskets inside wooden coffins and returned to Sweden with moving ceremonial. Royal warships and seaplanes escorted them into Stockholm, soldiers stood to attention and military bands played. Cremating the remains afterwards seemed an anticlimax.

As yet, no one asked the vital question: why didn't they return alive? What killed first Strindberg, whom the other two buried, and then Andrée and Fraenkel? Was it really just cold and exhaustion? And a final question: were both Andrée and Fraenkel really dead when the bears dismembered them?

Some suggested that the three men had died of starvation. But there was plenty of food in the camp, the stove to cook it with, thousands of matches to light it, and the means to get more food – a rifle, shotguns, plentiful ammunition. The authoritative version, put forward by Professor Nils Lithberg of the Scientific Commission, agreed with the quoted views of a witness who had been on the *Isbjörn*: 'I think they died in their sleep! – that the cold finished them!' The men's clothing was unsuitable for the Arctic, it was found, and the remains of their makeshift tent – under which, it was now known, Andrée and Fraenkel had died side by side and more or less simultaneously – were made of thin balloon fabric. They were frozen to death! That was the simple, official answer. A good, popular recipe too, but not good enough for some people, who began to search for clues in the diaries of the dead men, as soon as these had been published by the Scientific Commission.

Perhaps the most remarkable feat of preservation for the delicate material was the work done on Strindberg's films by J. Hertzberg. The expedition's films were Eastman-Kodak, marked for development by February 1898, after which date they carried no guarantee, even in perfect conditions. They were now to be processed in 1930, after being kept for thirty-three years at satisfactorily low temperatures, but soaked through with water instead of being stored in a dry place. Four long rolls of film containing 192 exposures were developed. Fifty of these negatives had some sort of image on them, of which twenty eventually gave reproducible pictures. (See plates 3–4). These photographs bore out the testimony of the diaries: they showed three proud, tough, determined men sledging and hauling their gear over all obstacles, confident of being able to win over the harsh Arctic environment.

While still airborne on the second day, 12 July, Andrée had written:

> It is not a little strange to be floating here above the Polar Sea. To be the first that have floated here in a balloon. How soon, I wonder, shall we have successors? Shall we be thought mad or will our example be followed? I cannot deny but that all three of us are dominated by a feeling of pride. We think we can well face death, having done what we have done. Is not the whole, perhaps, the expression of an extremely strong sense of individuality which cannot bear the thought of living and dying like a man in the ranks, forgotten by coming generations? Is this ambition?

The 'Eagle' had flown, on and off, from 12.46 hours on 11 July to 07.19 hours on 14 July, but had been airborne during that time only for 65 hours 33 minutes. Almost from the first, there had been a serious loss of gas. This was due to the fall of the car into the water immediately after the start. To avoid disaster, Andrée had dropped 455 lb of sand ballast, but this was too much and so the lightened balloon rose too high, to 1980 feet, causing the gas to expand and some of it to be valved off and lost. Then, when the balloon drifted out of sunshine and into cloud or through fog, the gas cooled and contracted and the balloon sank, and more ballast had to be jettisoned. The sphere went up and down like a rollercoaster, according to the varying lift given by the gas, rising as high as 2310 feet and at other times coming right down on the ice and bumping along, shaking everyone inside violently. Height and heat both caused the gas to expand; fog and low altitude were the facts that made it contract. Steering was impossible, and the 'Eagle' floated on varying courses. When finally, with limp envelope, it would fly no more, the balloon had covered only 288 miles in a direct line to the start point, and these in a northeasterly direction. That final crash position was almost exactly 200 miles due north of White Island, which lies just off the northeastern tip of Spitsbergen. It was very close to the crash position of the airship *Italia* in 1928, and nowhere near – mostly by many thousands of miles – the areas where witnesses claimed to have seen either wreckage, or a balloon in flight, or Andrée and his companions in a vision.

On the other hand, those Arctic experts who had predicted that the balloon party would try to sledge out by making for a food depot at Cape Flora on Franz Josef Land off the coast of Russia, were correct. What they did not allow for was the variable drift of the pack-ice, which moved faster under the force of wind or currents than a man could move across that terrible terrain. Strindberg wrote a letter to Anna Charlier on her birthday, 24 July:

> We have stopped for the night on an open place; round about there is ice, ice in every direction. Hummocks, walls, and fissures in the sea alternating with melted ice, everlastingly the same. For the moment it is snowing a little . . . We are moving onwards so slowly that perhaps we shall not reach Cape Flora this winter, but, like Nansen, we shall be obliged to pass the winter in a cellar in the earth. Poor little Anna, in what despair

you will be if we should not come home next autumn. And you can think that I am tortured by the thought of it, too, not for my own sake, for now I do not mind if I suffer hardships as long as I can come home at last.

Each man had a sledge, heavily loaded, which they had to pull or push across the jumbled icescape. Andrée's at first carried a load of 210 kilos (463.7lb). When they came to stretches of open water, they had to raft the sledges across on the canvas boat. Quite early on, Strindberg's sledge fell into the water, an incident he described in a letter to Anna: 'We managed to get the sledge up, but I expect that my sack which was on the sledge is wet inside. And it is there that I have all your letters and your portrait. Yes, they will be my dearest treasure during the winter.' It was a dreadful labour, but they got through by working out a routine to be followed each day, and with rough jokes: 'I don't need a wash today, I had one the day before yesterday!' Then of a bear they shot: that it was undoubtedly the oldest bear in the Arctic and had probably escaped from a menagerie, it was so tough. They celebrated anniversaries with formality: 'Fourfold hurra for N's sweetheart when the 25 July broke,' noted Andrée. They had a tent for protection, but only one sleeping bag; very little space, but only one case of serious dissension between them is noted.

As explorers, they made scientific and natural history observations, taking samples where possible and loading them on the sledges. They recorded a number of sightings of a 'monster' or 'sea serpent', some very large animal indeed seen in the distance. By 1 August, food was getting short. 'We are longing for bears for the meat is finished,' wrote Andrée. 'This evening we have seen the back of a new animal which looked like a long snake 10–12 metres (33–39 feet) long of a dirty yellow colour and, in my opinion with black stripes running from the back for some distance down the sides. It breathed heavily almost like a whale which I suppose it really was.'

On 4 August they decided to give up the attempt to walk eastward to Franz Josef Land and instead go south to the Seven Islands off the north coast of Spitsbergen, and winter there. On 7 August, Andrée noted that they all seemed to have a 'permanent catarrh' now; and next day there is the first mention of bowel trouble: Fraenkel has diarrhoea. By 15 August, Andrée and Strindberg are suffering similarly. But all of them have recovered by 3 September. On 21 August, they had made a change in their diet. Previously, they had cooked the meat of the bears they had killed, but that evening three bears came up to attack them as they were pitching their tent: a mother and two cubs. Strindberg put down the she-bear with one shot; Fraenkel got a cub with two shots; Andrée hit the other cub four times but did not bring it down, and the poor wounded creature escaped among the fissures and pools. By now, the meat-hungry men regarded bears as 'walking butcher's shops'.

'We took the best bits i.e., 2/3 of the tongue, the kidneys and the brains,' wrote Andrée. 'We also took the blood and F. was instructed to make blood-pancake (my proposal). He did this by using oatmeal and frying in butter after which it was eaten with butter and found to be quite excellent.' Probably it was their debilitating and depressing gut trouble which caused them to experiment. 'This evening on my proposal we tasted what raw meat was like,' wrote Andrée. 'Raw bear with salt tastes like oysters and we hardly wanted to fry it. Raw brain is also very good and the bear's meat was easily eaten raw.'

On 25 August they were still very sick, with pains in their limbs as well as the bowels. Nothing stopped their scientific observations, however. 'The sea-serpent was seen but looked different. He still appeared to have two curves but now he seemed to be grey everywhere and when he dived a two-cloven fin was seen at the end.'

They became more successful in dropping birds with a shotgun; once three ivory gulls were killed by a single shot. Andrée noted on 9 September that the diarrhoea had now stopped, and by then the mystery of the 'sea monster' had been solved, by a meeting at close quarters head-on. 'We turned back and met a walrus whose noise and behaviour in other ways and whose habitus (sic) showed us that we had met "the sea serpent". Consequently they were walruses we had heard and seen all the way up.'

Andrée made a lengthy entry on 17 September, for it was a momentous day. They had killed a seal two days before and were now all right for meat for three weeks. The gulls were tasty and easier to get, but took too much ammunition for the flesh on them. They had realised that the pack ice was drifting them rapidly south, defeating any attempt to make good a definite line of march to safety, and they had accepted the necessity of wintering on the ice.

Also they had seen a white mountain in the distance. Now, on this day, they recognised the land in front of them. 'It is undoubtedly New Iceland that we have had before our eyes,' wrote Andrée, drawing a rough sketch. It was White Island. 'There is no question of our attempting to go on shore for the entire island seems to be one single block of ice with a glacier border.' It was a forbidding prospect, but the drift from the frozen north southwards was welcome. 'Our humour is pretty good although joking and smiling are not of ordinary occurrence,' wrote Andrée of this period.

On Sunday, 19 September they had shot enough seals to last them into the New Year, probably to February, and Strindberg had begun to construct a house out of snow and water in which they could endure through the months of eternal darkness ahead of them. On 20 September Andrée noted for the first time 'signs of differences arising between us', hoping 'that this seed will not grow and develop.'

On 28 September their 'house' was ready and they moved in. In with them they took the bodies of the bears and seals they had killed, because if left outside other bears would eat them. While they were sleeping, on the early morning of 2 October, they 'heard a crash and thunder and water streamed into the hut.' The ice-floe which was to support them through the winter had fractured into many small pieces, and one wall of their hut overhung the water. Some of their provisions, including two dead bears, were now lying on a separate floe. These two represented rations for three or four months. At this point, that particular diary kept by Andrée ended.

The next book which he began suffered very badly. The first had been wrapped in senne grass (an Arctic insulator much used by Eskimoes) inside a woollen pullover. This one he had on him in an inside pocket of his jacket, together with the lead pencil he used for writing (ink would have frozen). Probably it had been covered by snow and then uncovered by a thaw, more than once. This would explain its pulpy, dilapidated condition and the mould sprouting from the glue of the spine. When the documents were published in 1930 (*Med Ornen mot Polen*, Bonniers, Stockholm) and 1931 (*The Andrée Diaries*, Bodley Head, London), very little of this memorandum book could be deciphered. Later, after further treatment, much more became readable, but the mystery remained. The words were enigmatic. The final date, however, was 8 October 1897. In Strindberg's much briefer notes, the final pencilled entry is 6 October (there is an ink entry for 17 October, probably made long before).

What had happened was that on 4 October the southward drift of the ice-floe had revealed an unsuspected low-lying part of White Island on which it might be possible to land, not merely by themselves but with their food and equipment. Next day, they succeeded in doing this, but only after such hard labour that they were still working after midnight and into the following day, which was the birthday of Andrée's mother. He was greatly attached to her, for his father had died when he was but sixteen years of age, and so he christened their new home Mina Andrée's Place.

This is a much more prosaic event than Professor Nils Lithberg's guess from the partial evidence which was all he had to work with in 1930. He wrote: 'Had the situation, from being "exciting", as Strindberg expresses himself, developed into one that is serious, so serious that it has made them lose heart?' There is no support for that thesis in the last diary entries. On 6 October, Strindberg noted tersely: 'Snowstorm; reconnoitring.' Andrée wrote more, to explain that the strong wind and snow prevented much work being done and so instead they had made a short tour of their new surroundings. He added: 'During the evening we started in the dark to build a snow-house and to carry our belongings to its neighbourhood.' On 7 October, Strindberg's usual brief note consisted of just one word: 'Moved.' Andrée explained in detail why he had wanted to move their camp – because their present site was likely, on the evidence of the snowstorm, to become completely snowed-up in winter.

On Friday, 8 October, there is no entry by Strindberg – he had written for the last time the day before. There is a shorter than usual piece by Andrée, which is his last:

> During the 8th the weather was bad and we had to keep to the tent all day. Still, we fetched enough driftwood so that we could lay the beams for the roof of our house. It feels fine to be able to sleep here on fast land as a contrast with the drifting ice out upon the ocean, where we constantly heard the cracking, grinding, and din. We shall have to gather driftwood and bones of whales, and shall have to do some moving around when the weather permits.

The inked entry in Strindberg's diary nine days afterwards misled everyone for a long time into believing that all three men were alive on that date. It was 1938 before Professor Hans W. Ahlmann of Stockholm pointed out that the entry was on a printed almanac page, being merely a note against a date: 'hem k1.7.5 f.m.' That is, 'Home 7.5 a.m.' The Professor suggested that this referred to the time of arrival of a train in Stockholm and represented the young man's optimistic appreciation, before he set out, of the probable date and time of his return to Sweden, and not a return from some hunting trip to 'Mina Andrée's Place', the last camp on White Island.

Other commentators found in the documents published by the Scientific Commission much material with which to combat the findings of that Commission, particularly the views of Professor Lithberg that the three men had died of cold. If so, it was odd that none of the men was found inside the large sleeping bag which they shared between them, odder still that there was a plentiful supply of spare clothing lying around unused, and oddest of all that there were heaps of driftwood to make a fire, paraffin to fuel it initially, and enough matches to last them a whole year! Not to mention food in plenty, including much bear, seal and bird meat. Further, the diary entries show that they were active and unworried until, suddenly, the entries ceased.

That this was so is shown most graphically by the typically brief, factual notes which Strindberg made during the first week of October:

2 S. Our ice-floe broke close to the snow-hut during the night
3 S. Exciting situation
4 M.
5 Tu. Moved to land
6 W. Snow-storm Reconnoitring
7 Th. Moving

Did death steal upon them unawares? If so, how? The Arctic explorer and writer Vilhjalmur Stefansson suspected that it had, and looked for causes which in his experience would be consistent with both the diary entries and the evidence of the artefacts at the camp. He pointed a finger of suspicion at the Primus stove found inside the tent area with the bodies of Andrée and Fraenkel. Carbon monoxide! A deadly, insidious gas inside a confined space, a favourite choice of suicides. He himself and his party had nearly died of it once. He had seen his companions grow drowsy and collapse, without knowing even that anything was happening to them, and had nearly drifted off into a final slumber himself.

Researching further, Stefansson found that he could also cite William Barents, who had wintered on Novaya Zemlya in 1596; his men in their ice house had also begun to faint when they made a great fire out of sea coals taken from their wrecked ship. And much more recently, the Byrd Antarctic expedition had nearly lost a night watchman who collapsed when a blizzard, unknown to him, stopped up the flue from the kerosene stove used for heating. Was this the fate that had overtaken Andrée and Fraenkel?

Stefansson thought so. The stumbling block was Strindberg, who had been carried away from the camp and buried between rocks under stones. As others before him had found, you had to explain Strindberg's death. You had to drown him, or make him fall off a cliff, or suffer fatal appendicitis, or be poisoned in some way, and all without a shred of evidence. But, except for the death of Strindberg, Stefansson's suggestions seemed plausible and were supported by Professor Yandell Henderson of Yale University and Dr H.U. Sverdrup of the universities of Oslo and Leipzig. His hope was that sufficient organic material might have survived the circumstances of time and the Stockholm disposal of the bodies for tests to be made to determine the presence of carbon monoxide.

Eventually, tests were made in order to substantiate an alternative theory, based on the plain evidence of the diaries that, although at first they had cooked the meat of the bears they killed (except for the liver, well known to be poisonous), from 21 August they had begun to eat bear meat raw and enjoy it, 'like oysters'. All of them had done this – Andrée, Strindberg and Fraenkel – and they had brought bear meat with them to their camp on White Island. A Danish doctor, E.A. Tryde, suspected that they had contacted trichinosis, an illness caused by eating uncooked meat containing the larvae of a certain worm. Sometimes there are no symptoms, but occasionally the effects are severe and may be fatal. In some conditions, bear meat has given rise to epidemics.

If this was the case, then the death of Strindberg a day or two before the others was explained. Now it fitted; there was no stumbling block. All had been exposed to the poisoning, and all had died of it very shortly after reaching White Island, for all the experts agreed that the camp had not been occupied long, indeed was unfinished.

Tests on the skin of a bear shot by the Andrée party proved conclusive – there were traces of trichinosis. The Polar bears had been more than 'walking butcher's shops', they had been a fatal, shambling infection as well.

Chapter 3

THE MONTROSE GHOST
The Irish Apparition: 1913–1916

The oldest ghost story in aviation and the most widely believed concerns a 'fey' little Irishman from County Clare called Desmond Arthur. Black-haired, grey-eyed, with a keen sense of personal honour, he was given to extremes of elation and depression. He gained his Royal Aero Club certificate on 18 June 1912 on a Bristol monoplane at Brooklands, the motor sport circuit and aerodrome south of London, and was killed on 27 May 1913 over Montrose in Scotland when the BE2 biplane he was flying folded up in the air at 2500 feet.

With the 70 hp Renault engine throttled back, the pilot had been gliding down from his original height of 4000 feet in a gentle turn. From the ground, all that could be seen was the result of the sudden collapse of the upper starboard wing. The yellow interplane struts came loose and the broken wing folded, throwing the little biplane into a series of convulsive, fluttering jerks. The uncontrollable gyrations snapped the pilot's seatbelt and hurled him from the fuselage. The watchers on the ground saw a dark object fall away from the fluttering wreckage, arms and legs working, and drop away beneath with ever-gathering speed until it hit the ground. No parachutes were worn in 1913.

Desmond Arthur would have had time to realise that every bone in his body was to be broken, that there was no escape from doom. All his friends could guess what his last moments had been like, and therefore their resentment was all the more bitter, even savage, when the Royal Aero Club's Accidents Investigation Committee found that the wing had collapsed because of a faulty repair carried out on the ground by a person or persons still unknown. And not only had a repair been done, but it had been unauthorised and afterwards concealed. In short, some guilty party had broken the wing, botched the repair, and then covered up the fact that the spar had ever been broken in the first place. And it had been done by some workman, some mechanic, who did not have to fly and therefore risk his own neck.

But there was no clue to identify the murderer and so help bring him to justice, for the wing could have been damaged in many ways and at more than one establishment. The damage was a fracture of the main rear spar slightly less than a foot from the wingtip. That could have been caused by a collision on the ground with a hangar doorpost or even with another aircraft; or by being dropped as it was being handled in workshops or while being fitted to the machine. This particular BE had been used over Salisbury Plain, then sent to the Royal Aircraft Factory at Farnborough to be rebuilt; from there it had been issued to the Royal Flying Corps at Jersey Brow, and from Jersey Brow flown by Major Charles Burke to Montrose.

When the spar had been fractured would never be known, for no one had logged either the damage or the repair. But for whoever flew the BE in that condition, it was a deathtrap. The break had been repaired with a 7½-inch-long taper splice so crudely made that the glue was only an eighth of an inch thick in places, and the whipcord binding it had not been varnished or treated with cobbler's wax. The guilty workman had then covered over the weak point in the wing with new fabric, and he was in such a hurry that he did not match the new piece to the original fabric on the wing. Desmond Arthur had not died in a flying accident. He had been murdered.

But there were no hauntings. His death had been both horrible and wanton, but the fey little Celt did not come back to haunt his murderer, nor even to revisit the scene of the crime done to him. Not, that is, until 1916.

With the war, everything had changed. Desmond Arthur had been a lieutenant in No. 2 Squadron, which had long since gone overseas. Montrose was now a training aerodrome occupied by No. 18 (Reserve) Squadron. Sited beside the North Sea, it had always been a lonely place, but the wartime expansion had brought a building programme to house the pupils, and it included a grand new officers mess. The instructional staff did not live there with the 'Huns' as the novices were called (because, it was said, they spent their time breaking British aeroplanes faster than the Germans could), but in the old mess, the original building used by No. 2 Squadron and known to Desmond Arthur.

One evening in the autumn of 1916 a senior member of the staff,

Major Cyril Foggin, was walking back to the old mess when he saw an officer in flying kit walking in front of him. The pilot walked up to the door but did not open it. When in his turn Major Foggin reached the door, it was closed and the other officer was not there. It was dark by then, but it was impossible that he could have stepped off the path and walked away without the Major noticing. Foggin afterwards tried to reason the apparition away as due to eyestrain or imagination, but inwardly he knew that he had seen somebody there, and that somebody had vanished in a way no human being could.

After a few days had passed, however, doubt crept in: Foggin began to accept that he must have been mistaken. Then one evening he saw the mysterious airman again, walking towards the old mess; once again, as the figure reached the door, it vanished. Two or three times after that, he saw the same apparition. The Major became seriously worried, but did not tell anyone else then what he had seen, or thought he had seen. Senior officers at flying training schools who see disappearing ghosts repeatedly are likely to be invalided as nerve cases or the victims of hallucination.

An uneasy atmosphere began to permeate the mess, because other officers also began seeing things which were not there – or simply sensing them. And at first all were too wary of ridicule to tell anyone else.

One instructor was soundly asleep after a hard day's flying, trundling 'Huns' round the sky, when he woke abruptly with a feeling that there was somebody in the room with him. He propped himself up in bed and by the dim light of the fire saw a man sitting in a chair at the foot of his bed. He asked the chap who he was and what he wanted, but there was no reply. The instructor leant forward to get a good look at the man's face. And then there was no one there. Had he been the victim of a particularly vivid dream which had continued into a half-waking state? The instructor felt he had not, but the basis for ridicule if he told of his experience was obvious, and he kept quiet at first.

Then, a little after that, two officers who were sharing the same room were woken up one night by a feeling of oppression. It seemed that while still asleep they sensed the presence of a third person in the room with them. The room was completely dark, with not even the embers of a fire flickering in the grate, so that they could not have seen anything. But one of the officers who had woken up uneasily spoke to the other in order to wake him up, too – and discovered that he was already awake, because he also had sensed a presence. One of them fumbled for a match, but when he struck a light the room was empty. For the moment, they told no one else of their experience although a shared hallucination might be regarded as more substantial than just one man's aberration.

But soon after a few other officers experienced much the same as these two. They were disturbed by the presence of someone or something which they could sense strongly but neither see nor touch. Always, the occurrences took place in the old mess, the former home of the pilots of No. 2 Squadron to which Desmond Arthur had belonged, and not in the more splendid building erected later to house the wartime pupils.

It was never established exactly how these stories suddenly became common property. Presumably one of the witnesses blurted out a tale of something very odd he thought he had experienced, and found at that instant that half a dozen others in the mess had had exactly the same thing, or something very like it, happen to them.

But the story did not stay at Montrose. Perhaps spread by the pupils as they finished their training and were posted to squadrons, the story of the Ghost of Montrose quickly went round the whole of the Royal Flying Corps, at home and overseas. C.G. Grey, the forthright and controversial character who edited a weekly technical magazine called *The Aeroplane*, heard of it and discussed what had happened with the chief witnesses. The prewar RFC had been a small, elite body and C.G. Grey had not merely known Desmond Arthur but had been a personal friend of the little Irishman, whom he summed up as having been a 'singularly lovable person, though distinctly weird and, in the current phase, "psychic"'.

Of course, as the story spread it was presented in any number of ways according to the beliefs of the individual telling the tale or his judgement of the mood of his audience. For some, it was just another ghost story to be taken with the brandy at Christmas, a mild, momentary flesh-creeper. For others, it was an appalling insight into the state of nerves among the wartime instructional staff at home, a matter for contempt or concern, according to the views of the teller. There were a few who took it seriously and C.G. Grey was one of them.

Up to this time there had been no mention of Desmond Arthur. No one had given the ghost a personality, let alone pinned a name on to it. The story was just of some uncanny happenings at Montrose in 1916.

There was one attempted explanation, however. It is surprising there were not more, because it is tempting to rationalise a haunting, to find a reason, and the most satisfying seems to be to postulate a victim of some sort. It was suggested that the disturbing presence was that of a 'Hun' who had been told by his instructor that he was now ready for his first solo – always a traumatic moment, although a first solo is often the safest flight a pilot ever makes, because he is so terrified that he pays attention to what he is doing and does not attempt to show off. In this case, went the story, the pupil told his instructor that he did not feel confident and did not want to fly, but the instructor had insisted. The 'Hun' had taken off, crashed, and killed himself.

C.G. Grey did not like this story as an attempted explanation for the Montrose Ghost. First of all, if every pupil who crashed fatally in attempting to carry out his instructor's orders had come back to haunt him, wartime aerodromes would be full of apparitions. Secondly, this ghost never appeared in the new mess which belonged to the pupils; not only did it confine itself to the old mess of No. 2 Squadron, but it acted as if it was at home there, according to all the accounts which C.G. Grey had collected.

I can confirm this latter point, although only at third hand. My grandfather was stationed at Farnborough, the home of British military aviation, from 1911 to 1914, as Captain and then Major in the Royal Munster Fusiliers. No. 2 Squadron was at Farnborough until it flew up to Montrose in 1913, and, like C.G. Grey, my mother knew Desmond Arthur personally. So when my mother spoke of Desmond Arthur I listened, and the story I heard of the Montrose Ghost was that it was most often seen sitting down in an armchair in the mess, reading, apparently oblivious to anyone else. That description became fixed in my mind when I was still very young as the essential image of the Ghost. I was told also of the broken wing, the criminally botched repair, the mid-air break-up, and the long fall to earth of the little Irishman. It always made me angry to think of it.

The point that C.G. Grey made in arguing that the apparition was the ghost of Desmond Arthur was to ask why, if Lieutenant

Arthur was killed by someone's negligence in 1913, did he wait until 1916 before turning up in his old mess as a disturbing presence? And the answer at which Mr Grey arrived was that while in 1913 someone had merely killed Lieutenant Arthur, in 1916 a group of people acting for the government had officially but untruthfully blamed him for the crash and so stained his personal honour. He had therefore turned up in his old quarters at Montrose, among his friends, in the hope that they might act to right the wrong done to him, for they certainly knew the truth.

The facts regarding the broken wing had indeed been established by the Royal Aero Club's Committee, and in very great detail, shortly after the crash. And of course everyone in the RFC knew what had happened. The public, however, did not, and could therefore be manipulated if it was in the interests of authority to do so. Further, while the Royal Aero Club was an independent body, governments can set up semi-captive committees.

C.G. Grey pointed to the significance of the dates. His careful questioning of witnesses showed that the 'Ghost' had first been seen in the autumn of 1916 and its last known appearance had been in January 1917. What, he asked, had disturbed it? Were there any related events reported from precisely that span of time?

Indeed there were, as he well knew, because he had been concerned in them. The government had been having an unhappy time in the spring of 1916 with the so-called 'Murder Charges' being laid against it by Mr Pemberton Billing, MP, backed by Lord Montagu and Sir W. Joynson-Hicks. The MP was the head of Pemberton Billing Ltd of the Supermarine Aviation Works at Southampton, the firm which later designed the Spitfire. With all the authority of a leading aircraft manufacturer, he called for a judicial inquiry into both the military and naval wings of the air service, the RFC and the RNAS, on the grounds that 'certain officers had been murdered rather than killed by the carelessness, incompetence or ignorance of their senior officers or of the technical side of those two services.' The government denied an inquiry into the RNAS, where their defence was paper-thin, but decided to fight over the RFC.

As attack is the best means of defence they went all-out to clobber Pemberton Billing by proving him a liar. They set up a semi-captive committee of their own, which on 3 August 1916 issued an interim report establishing among other things, that there was no truth in what he had alleged in the case of Desmond Arthur. The final item in their Report was calculated to damn Mr Pemberton Billing for ever. It read:

> STATEMENT BY MR PEMBERTON BILLING. – Desmond Arthur was killed on some type of BE machine which had been repaired by the Royal Aircraft Factory. The repaired part broke at 4000 feet up, and the pilot was pitched out.
>
> SUGGESTION. – Faulty design or bad repair.
>
> FACTS. – The date was the 28th [sic] May, 1913. The place Montrose. There was a suggestion made at the time that there had been a patch on the outside of the right wing of the plane, and that someone had broken the tip of the wing, then repaired it, and put a patch on the repaired part. The suggestion being that this was done by someone with a view to hiding some damage which he had done to the machine. The matter was closely enquired into at the time by a Committee, of which Mr H.T. Baker, MP, was Chairman. The Committee have had the notes of the whole of the evidence given to that Committee before them. There were some 23 witnesses. The suggestion depends on the unsupported evidence of one man out of these 23 witnesses. No useful purpose would be served in re-opening the matter, especially as some of the witnesses called have since been killed.
>
> A perusal of the transcript of the notes of evidence leads to the conclusion that the suggestion of the patch is quite unfounded.

After that, Mr Pemberton Billing could consider himself well and truly clobbered, and Desmond Arthur would be turning in his grave, for if there had indeed been no botched repair of a broken wing, then he must have broken the BE himself by foolishly dangerous flying. The hauntings did not begin immediately, however, but in the autumn, which was the time when the Final Report of the committee was being prepared (it went to the printers in October, was dated 17 November, but was not published until Christmas). This was precisely the period when the ghost was seen or sensed at Montrose. It was also the period when his friends could most usefully bring pressure to bear on the committee or at any event on public opinion. The reason was that two members of the Inquiry Committee had not only learned of the existence of the Royal Aero Club's report, but had insisted on referring to its contents in an addendum to the Final Report of their own committee. They were Sir Charles Bright, an engineer, and Mr Butcher, KC, a lawyer. The 'Further Memorandum from Mr Bright and Mr Butcher' included the following, a total contradiction of their colleagues' findings:

> Since the issue of our Interim Report attention has been called to the finding of the Royal Aero Club's Public Safety and Accidents Investigation Committee regarding the fatal accident to Lieutenant Desmond Arthur. It is to be regretted that the said Committee's Report – and the evidence on which it was founded – was not brought to our notice by any witness dealing with the subject during the course of the enquiry . . . It appears probable that the machine had been damaged accidentally, and that the man (or men) responsible for the damage had repaired it as best he (or they) could to evade detection and punishment.
>
> In making this statement we are taking the first suitable opportunity of amending – so far as we personally are concerned, and to the extent indicated – what appears on pages 7–8 of the Interim Report.

C.G. Grey remarked pointedly (in *The Aeroplane*, 29 December 1920): 'It has always puzzled me to understand why these two were the only members of the Enquiry Committee to act in this way.'

The damning Final Report of the semi-captive committee having been itself internally damned by the Bright and Butcher Memorandum, Desmond Arthur could rest, vindicated. He had been the victim of a crime, not of his own folly. After a final appearance in January 1917, the ghost was seen no more.

That is the true story of the Montrose Ghost according to the original evidence of some sixty or more years ago and with C.G. Grey's interpretation of its meaning at the time. Whether or not one believes in the ghost is a matter of individual judgement. Since then the story has been rewritten as a fiction and given a compelling dramatic structure. In this version, a modern pilot flying a modern light aeroplane gives a lift at Montrose aerodrome to an airman who has been stranded there and then, while in flight, sights the old BE2 of 1913 in the air and sees the wing collapse and Desmond Arthur thrown out with an awful shriek. Then he finds himself alone – his passenger has vanished! So he recalls that this day is the 50th anniversary of the tragedy. Undoubtedly this makes for a strong ghost story and a more colourful ghost. But I shall always remember the strange tale as I first heard it, with the likeness of Desmond Arthur sitting quietly, obliviously in a mess chair years after his death.

One thing is quite certain. There were no hauntings at Montrose after 1917, although it continued to be a training aerodrome for many years. In 1973 a revived interest in the story brought a comment from G.M. Macintosh in the pages of *Flight*:

> I instructed at Montrose during 1937–9 but I never saw the ghost. I wish I had. But it was the sort of place where odd things happened. I finished one evening's night flying as Paraffin Pete. The last Solo (an Oxford trainer) was nowhere to be seen or heard. I recall the awful, ghastly silence after the last engine had been switched off, broken only by the faint lap of tiny wavelets on the shore and the wail of waking sea birds. And nothing was ever found, nothing ever washed up, of that Oxford – nothing. We started an air search as dawn broke but all we saw were the mountains and the North Sea. Nothing to do with the Ghost of course but it was – eerie.

Chapter 4

WHO KILLED THE ACES?

Immelmann, 1916; Ball, Guynemer, 1917; Richthofen, Luke, 1918

One of the oldest aviation books I possess is a curiosity from the Great War. The title page reads: 'THE RED AIR FIGHTER, by Manfred Freiherr von Richthofen, with Preface and Explanatory Notes by C.G. Grey (Editor of *The Aeroplane*), London, 1918'. It cost 2/6d then, and the name on the flyleaf shows that it belonged to my uncle at one time. What is so extraordinary is that this book, a translation of Richthofen's wartime memoirs, was published in England during the war. Such a proceeding would have been impossible during the barbarous affair that took place between 1939 and 1945. No doubt, by the time it was published Richthofen was dead – shot down (I used to believe) by Captain Roy Brown flying a Sopwith Camel, although now I'm not so sure.

Indeed, it is curious that so many of the most famous of the 'aces' of the Great War should have died in mysterious or even violently controversial circumstances. There was Immelmann, a German ace from the very early days, in 1916; Guynemer the Frenchman and Ball the Englishman, both in 1917; Baron von Richthofen and Luke the American, both in 1918.

The case of Immelmann was odd in another way. So many authors on both sides of the Atlantic apparently started by mistaking him for someone else. The person they described has no connection with the real one. They begin by having him fly over Paris in August 1914, dropping a message to the Parisians threatening doom if they do not surrender instantly to German arms. Later they have him and Captain Ball exchanging duelling notes across the lines. And, of course, he must be an aristocrat of the military caste, in this case a Count of Westphalia.

In fact Max Immelmann was born at Dresden, a gracious city in Saxony, hundreds of miles from Westphalia. His father owned a cardboard factory, his mother was an auditor's daughter. But his father died young, his mother was stricken by illness, and the three youngsters of the Immelmann family had a hard childhood. Max found no pleasure in playing with the toy soldiers his younger brother collected; indeed he barred them from the extensive model railway systems he had laid out in his room. Later, he graduated to taking motorbikes to pieces, excelling in the more difficult science of putting them together again correctly. The family were brought up under a nature cure regime which Max thoroughly believed in: no meat, no tobacco, no alcohol.

His grandfather advised that he be sent to the Dresden Cadet School, as there were advantages to this in a country like Germany, where a period of military service was compulsory. He put up with it until he was seventeen, then told his mother that he did not fancy an army officer's career because he did not want to be bound by the stiff etiquette of the military caste; he preferred some technical profession. Reluctantly, he settled for the railway branch of the army, as this would give him the assured future his mother wanted for him, although there was less scope for his technical ambitions than he had anticipated. However, once he had gone through the required courses, he was able to resign with honour and enlist at the Technical High School in Dresden.

In August 1914 the coming of war altered everything. On 18 August his call-up papers arrived and next day he rejoined his old Railway Regiment. On 21 August, far from flying over Paris dropping bombastic, militaristic leaflets, Immelmann was writing of his life as a lance-sergeant in the army: 'Service here is idiotically dull. I am near to my spiritual death.' He had applied to join the Aviation Corps, but there was a long waiting list, and it was November before he was accepted for flying training. In December 1914 he was writing home:

> I have now been in the air twenty times. Unfortunately, I do not notice any increase in my ability. The business is not at all dangerous and not so hard as I imagined, but you have to look out when you land. Flying is splendid. You never feel it's not safe. It is far more exciting than in a car. The glorious peace in the air – and – no policeman!

In the spring, he began war flying and acquired a military haircut which, he complained, made his head look like a 'lavatory brush'.

The Germans were on the defensive in the air. They had begun by overestimating the efficacy of the giant Zeppelin and underrating the aeroplane, whereas the French had made an accurate evaluation and were on top. Their aircraft were usually armed with machineguns, not pistols or rifles. They did not confine themselves to tactical cooperation with the army; they carried out bombing raids into Germany, against cities such as Cologne, Friedrichshafen (where Zeppelins were built), Freiburg and Düsseldorf. Further, they had begun to group three or four 'scout-type' machines together, under a leader, and use them to attack German aircraft.

Air units were not uniformly equipped at this time, on either side of the lines. Different classes of machine were employed within the squadrons. The German 'B' category aircraft were unarmed tactical reconnaissance machines, as helpless as the BE2cs of the RFC. The 'C' category were also two-seaters, but the pilot had been moved to the front cockpit and the observer installed in the rear seat with a movable machinegun. This type could not fire forward but only in certain arcs to the rear, but at least it could defend itself, and it was used for long reconnaissances behind enemy lines. Some attempts were made to employ the type offensively, so it was called the 'fighter machine'. A larger type of 'battle plane', with twin engines and two machineguns, was used on 'barrage patrols' to stop enemy air incursions and to protect captive balloons, but was so clumsy and vulnerable that it was soon relegated to night bombing.

The machine of the future, which began to reach the front in spring 1915, soon after Immelmann had begun flying two-seaters, was the 'E' type. 'E' stood for 'Eindecker', or monoplane, and specifically it was a small, light single-seater designed by the Dutchman Anthony Fokker, and powered by the Oberursel rotary engine, first of 80 hp, then 100 hp, finally 160 hp. Outwardly, it looked much like the French Morane-Saulnier monoplanes, but the fuselage was of modern welded steel tubing although manoeuvre was not by hinged control-surfaces but by wing-warping, as old as the Wright brothers.

On 19 April of that year, 1915, a Morane-Saulnier flown by the French ace Roland Garros force-landed behind the German lines near Courtrai. Garros had brought down five German aeroplanes in sixteen days and now the Germans were to find out how he had done it. The answer was a fixed machinegun firing forward. But why didn't it also shoot the propeller to pieces? There was the answer, bolted to the propeller blades – discoloured steel wedges. They were deflector plates, a crude but effective method of turning an aeroplane into a killing machine.

The Germans invited Fokker, the designer of a similar aeroplane, to look at this novel French idea. Fokker did not copy it, because a far superior method had already been patented two years earlier (15 July 1913) by a Swiss engineer, Franz Schneider, who was technical director of the German aircraft company LVG. Two members of Fokker's staff knew about it. In this design, a device of cranks and rods – an interrupter gear – allowed the gun to fire whenever the propeller was horizontal, but prevented it from firing when the blades were vertical. It was neater than the deflector plates which sent ricochets whining off the steel wedges, and safer, but not, as it turned out, completely safe.

Equipped with a Parabellum machinegun and the new designation E I, Fokker monoplanes began to reach the front in small batches. Two of the first pilots to get a chance at flying it were Oswald Boelcke and Max Immelmann (who up to now had been flying an LVG biplane and was not qualified to fly monoplanes). On 1 August 1915, Immelmann had had two days of cautious practice on a Fokker E I, but had only fired thirty rounds at a ground target, when at dawn about ten enemy aircraft arrived over the German field at Douai and began bombing it. Boelcke took off, but suffered a jammed gun. When he returned, he found that Immelmann had hopped into the reserve Fokker and taken off. He was most perturbed, shouting out: 'They will shoot our Immelmann dead!'

Immelmann got above a single enemy machine, but with two more of the enemy above him in turn; a dangerous situation. He closed to fifty yards and fired. After sixty rounds, his gun jammed. He had to take both hands off the controls to clear the jam. The gun jammed twice more, while he manoeuvred to head the enemy off. After ten minutes he and his opponent were down to about 1200 feet and the other two enemy were rattling away at him with their guns. After firing nearly 500 rounds, Immelmann's gun stopped for good, but the plane he had been firing at lost height steeply in a glide. It landed and he landed beside it, and only then realised that he was unarmed and the countryside deserted.

But there was only the pilot in the machine. The gunner's cockpit was empty (he had been left on the ground in favour of a heavier bomb-load), and the pilot was helpless, badly wounded in one arm. His name was William Reid. Immelmann helped him out, then, when the first cars arrived, sent one for a doctor. Then he turned to examine the results of his own fire. About forty hits, he judged. Two in the propeller, but none in the engine; three in the petrol tank; more in fuselage and wings, and all the instruments shot to pieces. There was a bombsight fitted but no machinegun. Reid had deliberately come over unarmed.

Nevertheless, what Immelmann had done instinctively was far removed from the existing German conception of a 'barrage' defence. He had gone out and hunted his enemy down. This was the concept which fitted the new Fokkers best, but it was dangerous because there were so few of them that they had to operate like lone birds of prey, covering a great area of front. Of the first fifteen Fokker pilots of 1915, none lived to see 1917, although the machines improved. The E I had a top speed of only 81 mph. The E II with a more powerful engine could touch 87 mph at sea level. Then a second machinegun was added, in the E III. But the E IV, with two rotary engines in tandem and three machineguns, was a failure. Even so, this handful of underpowered, unreliable, rickety monoplanes gave Germany air superiority for nearly a year – the year of the so-called 'Fokker Scourge'.

The introduction of more than one machinegun led to alarming failures with the interrupter gear. In March 1916, Immelmann shot his own propeller completely off – both blades simultaneously sawn through – and had to hurriedly force-land. Soon after, Boelcke shot off a single blade of his propeller, and this proved to be more serious, because it set up an irregular strain which broke the struts connecting the engine to the fuselage. Still, he survived. As did Immelmann on 31 May when he similarly shot away one propeller blade. In this case, he had fired a very long burst at extreme range to distract and scare two British airmen who were about to surprise a novice Fokker pilot from above.

Instantly Immelmann's Fokker reared up, and began to shake and jolt. He switched off petrol and ignition, but a rotary engine, in which the cylinders revolve, takes some time to lose momentum. The fourteen cylinders went whirling round, continuing to drive half a propeller; weird tremors went through the machine, then the engine finally stopped with a violent jerk. The monoplane flipped over on its left wing and hurtled downwards.

The position of the unbroken blade of the propeller was odd, and there was only one explanation. The struts holding it to the fuselage must be fractured, and at any moment it might fall away. From behind there came curious creaking sounds. A rapid glance over his shoulder showed Immelmann that the rudder was flapping idly in the wind; moving his feet on the rudder bar confirmed the impression – it was like kicking cottonwool. The Fokker was going earthwards very steeply and the patchwork pattern of fields was expanding dangerously fast.

Cautiously, Immelmann tried the stick now, and could sense welcome pressure through the elevator wires. He eased it back further, and as the nose came up, the engine seemed to slip back with a jerk, nearly into place. That was better, but he dared not pull back too hard or the wings might sheer off. Gently, he eased the stick still further back. At about 1500 feet the crippled machine checked its plunge and Immelmann jockeyed it down in rollercoaster fashion, trying to keep the loose engine from breaking away completely and compelled to glide straight ahead. But would the Fokker flatten out for a landing? Immelmann had hardly time to think about it before he was down in a meadow alongside the Cambrai-Douai road, the engine loosely attached by two bent struts and with some of the bracing wires entangled in the rudder control wires.

Apart from incidents such this, suffered by both Immelmann and Boelcke, engine failure was common and gun stoppages routine. Nevertheless, Immelmann managed to run up a score of sixteen enemy aircraft shot down, confirmed by witnesses, and became a national hero. Anthony Fokker, who knew him well, described Immelmann as 'a serious, modest youngster intensely interested in the technical details of flying'. In his letters home, he did not glorify himself. In one, he wrote: 'I have already told you about the airfights in which I got the worst of it – twice in the biplane and three times in the Fokker.' He reported almost every evening to Major Stempel, Aviation Staff Officer of Sixth Army, so that improved fighting instructions could be worked out. Above all, he advocated the grouping of the single-seater fighters into squadrons and their use en masse rather than doled out in ones and twos.

On 10 June, Immelmann was ordered to form the first *Jagdstaffel*, or hunting squadron. On 18 June the unit was still being got together, and had not yet received the first of the new Halberstadt biplanes which were to replace the now obsolescent Fokker Eindeckers, when a British squadron of eight machines was reported crossing the lines. Immelmann led a force of four Fokkers to intercept, but he was the only German to score. In the evening seven British aircraft were reported crossing, and were soon engaged by the late standing patrol of two Fokkers flown by Lieutenants Mulzer and Osterreicher. They were reinforced by a third machine flown by Sergeant Prehn, who took off just as Immelmann and Corporal Heinemann, called away from their evening meal, arrived at their machines. Immelmann was last away, and climbed towards an untidy situation.

In the northeastern part of the sky two of his Fokkers were attacking four British aircraft, while in another part a second pair of Fokkers were climbing hard towards three British machines, all five being fired at by the German anti-aircraft guns. Immelmann screwed his Fokker up towards this group, firing white flares as a sign for the German guns that they were endangering German machines. Then, gaining more height, he dived on one of the three British machines. As he did so, another British aircraft left the northeastern dogfight and came at him, itself followed by another Fokker flown by Lieutenant Mulzer, whose fuel was almost exhausted. Sergeant Prehn and Corporal Heinemann left their opponents to shield Immelmann from the newcomer.

Heinemann saw Immelmann's victim fall away earthwards, followed by Immelmann, followed by Mulzer (who was going to shepherd it down, then land from lack of fuel). The British machine, which the Germans believed to be a Vickers, landed intact near Lens, because the pilot had been wounded and was losing blood fast. Heinemann saw Immelmann turn away from his last victim to engage another enemy machine which was about 3000 feet away. While climbing to gain height, Immelmann's Fokker reared up in curious fashion, then did a diving turn to the left, but levelled out and flew ahead with odd lashing movements of the tail. Three British biplanes came roaring down into the fight, picking on Sergeant Prehn. Heinemann had to bank away to help the Sergeant and lost sight of his Staffel-leader.

Watchers on the ground followed the strange gyrations of the

Fokker with horror. At about 6000 feet the fuselage broke in two just behind the cockpit and was left fluttering on high. The front end containing the heavy engine and the man fell faster and faster with a weird whistling sound. The wings collapsed and tore away, and were left trailing above. The nose of the Fokker with the rotary engine and the pilot now came down like a stone and hit the ground with a dull, thudding explosion. The first soldiers to run to the crash opened the dead pilot's leather jacket. They could see the rare 'Pour le Mérite' decoration, the 'Blue Max'. On the linen the monogram 'M.I.' – for Max Immelmann.

The two sections of fuselage were found several hundred yards apart, the metal longerons of the framework showing clean fractures as if blown in two. Many witnesses had seen the Fokker break up in the air, and so the rumour went round: 'Our Immelmann has been shot down by a direct hit from an anti-aircraft gun!'

That was not what the pilots themselves were saying. They all knew of the narrow shaves experienced by both Boelcke and Immelmann; they thought they recognised the frenzied oscillations of a Fokker which had shot away part of its own propeller. They concluded that it was the Fokker's interrupter gear which was at fault.

These stories reached Anthony Fokker and alarmed him. He was to write:

> Almost as much mystery surrounds the manner of Immelmann's death as Guynemer's, which was never adequately explained . . . It was first given out that his [Immelmann's] Fokker fighter had failed in mid-air. This explanation naturally did not satisfy me, and I insisted on examining the remains of the wreck, and establishing the facts of his death. What I saw convinced me and others that the fuselage had been shot in two by shrapnel fire. The control wires were cut as by shrapnel, the severed ends bent in, not stretched as they would have been in an ordinary crash . . . There was a strong opinion in the air force that his still comparatively unknown monoplane – which somewhat resembled a Morane-Saulnier – had been mistaken for a French plane. I was finally able to convince air headquarters sufficiently so that, while it was not stated that he had been shot down by German artillery – which would have horrified his millions of admirers – neither was the disaster

blamed on the weakness of his Fokker plane. The air corps exonerated the Fokker plane unofficially, although as far as the public was concerned the whole incident was hushed up. Because of this investigation, however, silhouettes of all German types were sent to all artillery commanders to prevent a repetition of the Immelmann catastrophe.

What Anthony Fokker did not add was that the interrupter gear was also altered, so that gun and engine were truly synchronised; in future, no Fokker could shoot its own propeller off. The designer was only too well aware of this habit, for he had suffered it himself while demonstrating the three-gun 160 hp Fokker E IV, which could fire 1800 rounds a minute. One gun went wrong, but Fokker felt the vibration in time, stopped firing and landed the monoplane at the far end of the field, where no one could get a close look at it and see that one blade of the propeller, with sixteen bullet holes in it, was ready to drop off. This machine was later given to Immelmann, who, as we have seen, met the same trouble which Fokker had been unable to cure. However, the machine in which he was killed was a twin-gun E III, the E IV having been too shot up in the earlier fight on 18 June.

Although there were conflicting accounts of how Immelmann had died, officially the British were in no doubt as to what had happened: one of their aircraft, a Farnborough design, flown by Lieutenant McCubbin, had shot him down in air combat. The Air Board announced:

> On June 18 one of our F.E. aeroplanes whilst patrolling over Annay at about 9 p.m. attacked three Fokkers. One immediately retired whilst the other two turned towards Lens and proceeded to attack another F.E. which was then approaching from that direction. The first-mentioned F.E. (pilot, Lieut. McC.; observer, Corporal W.) followed and joined in the fray, and diving steeply on one of the attacking Fokkers caused it to plunge perpendicularly into the ground. It was seen to fall to earth by one of our anti-aircraft batteries. A subsequent report from another machine in the neighbourhood states that the Fokker went to pieces in the air and both wings broke off. Extracts from the German newspapers relating to the death of Lieut. Immelmann make it clear that the pilot received his death as outlined above.

Boelcke, who had been in rivalry with Immelmann and flew over for his funeral, would not give this idea houseroom. He wrote: 'Immelmann lost his life by a silly chance. All that is written in the papers about a fight in the air, etc., is rot. A bit of his propeller flew off; the jarring tore the bracing wires connecting with the fuselage, and then that broke away.' Almost certainly, this is the truth of the matter. Fokker's explanation, and also the British official claim, although in rivalry, both represent wishful thinking.

However, it was true that, as Fokker had written in the 1930s, Guynemer's death was even more of a mystery than Immelmann's, and never adequately explained.

Georges Guynemer was born in Paris in 1894, so he was four years younger than Immelmann. Unlike him, he was born into a military family. Whereas the German was strong and stocky (Immelmann could give an almost professional acrobatic turn at unit theatricals), the French boy was frail and delicate. Immelmann wanted a technical career, Guynemer wanted to fly (but his father insisted that he go into banking). For Guynemer as for Immelmann, August 1914 changed everything, but while Immelmann had to wait his turn in the aviation queue, Guynemer was rejected five times as physically unfit. He would not give up, and became a student pilot in January 1915; by May he was flying the Morane-Saulnier two-seater monoplanes called Parasols. Immelmann had begun two months earlier, on the other side of the lines, and soloed on 31 January. While Immelmann, because of a single flying accident, got a not altogether deserved reputation with some of his foreign biographers as a bad landing artist, Guynemer really did break three aeroplanes in twelve days soon after joining his first front-line unit. But on 19 July 1915 he scored his first victory and was promoted sergeant, equal in rank to Immelmann.

In these early days, victories were few and far between, largely because opportunity was lacking, but partly because the aeroplanes were improvised war machines. However, even on the old two-seater Guynemer managed to destroy a Fokker in December 1915. Early in 1916, the escadrille (roughly the equivalent of a British squadron or a German Staffel) was re-equipped with new single-seat fighters, the Nieuport XI 'Bébé', a biplane with a Lewis gun mounted above the upper wing so that the bullets cleared the propeller arc completely. With this excellent machine the unit entered the battles of Verdun and the Somme, the major bloodbaths of the war. Guynemer's tactics were to dive straight in at a two-seater 'C' machine and trust to more speedy and accurate shooting to get the observer before the observer got him. As German gunners usually shot well, this was dangerous, and after only two days over Verdun, almost blinded by his own blood, Guynemer staggered back to his aerodrome with two bullets in his left arm and face and head pitted with fragments shot off his own windscreen. Three months later, he was back at the front with an even better machine, the Spad, which, luckily for him, was much stronger than the Nieuport.

This time he was actually shot down, engine out of action from a holed water tank, upper left wing stripped of all fabric. He had been hit by a shell from a French 75-mm anti-aircraft battery. The wrecked fighter spun instantly, but Guynemer eased it out into a dive before he hit the ground at a trifle over 100 mph (160–180 kph). The impact broke the Spad 'like matches', the wreckage bouncing into the air, flailing round forty-five degrees, then turning on its back and planting its remains in the ground 'like a post' forty yards away. About the only thing unbroken was Guynemer. He had bruised his knee.

One thing set the airmen aside from the trench soldiers. They returned straight from experiences like these to safety, normality, peace, cups and saucers, beds, sheets and blankets. And they often did this, several times a day. The contrast provided extra shock. There is an other-worldly look about their faces, particularly around the eyes, which shows distinctly in the photographs of the time; and it shows in the face of Guynemer as much as in Boelcke's and Ball's and Richthofen's. It is the sign of what was called 'Battle Exhaustion' in the Second World War, recognised not so much as a malady but as a definite progression based on experience. In World War II also, photographs of front-line soldiers capture that same haggard, dark-eyed look, of seeing beyond the camera into infinite distance. It is the gaze of the man who knows he is doomed.

In the Great War, the state was sometimes called 'Shell Shock' and not infrequently regarded as being akin to cowardice, which is grossly inaccurate. In the Second World War it was studied and graphed; Lord Moran, Winston Churchill's physician, even wrote a book about it called *The Anatomy of Courage*, based on his own observations in the trenches of 1914–1918. To simplify, the novice in battle, though nervous, has plenty of courage – and he believes himself personally bulletproof. But as he knows so little, he is hopelessly inefficient and vulnerable, very likely to be killed, in fact, and without first having paid for the expense of training him. But if he survives long enough to gain experience, he reaches optimum efficiency and optimum safety: he is at his most dangerous for the enemy, and wary enough to be very hard to kill. If he then proceeds beyond this point, and particularly if he is wounded or badly shaken up and so brought violently to realise that he is not invincible after all, he uses up the limited reserve of nerves and courage which he has. He knows that if he goes on long enough, death is certain. The usual reaction is for him to become very reckless at this point, but also very ineffective. He is literally too tired, too worn out, to win any longer, although his battle skills may by now be superb.

Immelmann was killed, almost certainly in an accident, before he reached this point. Boelcke was saved from it for a while longer by being sent to tour the Turkish front as a device by his superiors to preserve his life and give him a rest. Ball's superiors dealt with him in the same way. Richthofen was granted a good deal of leave, which had the same effect of prolonging his career out of all proportion. Guynemer's superiors, and the pilot's own parents, tried to get him to take up an appointment as an instructor of aerial fighting, for which he was superbly qualified. The concept of an operational tour consisting of a fixed number of missions, followed by a non-operational posting, was then a policy of the future. Guynemer, sleepless, strained, keyed up, angrily rejected the suggestion as an insult. He told his father: 'They would say I had stopped fighting because I had all the decorations France can give me.'

If anyone pointed out that he was worth more to France as a symbol, alive, than as a defeated hero, dead, then that argument did not work; but it would have been valid. The death of Immelmann, who had become a national hero, naturally had had a depressive effect on public morale in Germany, and it was bound to be the same with them all. The public tended to regard them as invincible champions, which was flattering but obtuse. In war, all too often, chance tips the balance or decides outright. Nevertheless, no one scored fifty-three victories, as Guynemer did, without being a deadly opponent. Oddly enough, there exist two personal testimonies to his skill, one written by a British pilot, the other by a

German. The one fought a mock duel with Guynemer – and lost it; the other fought a duel for real – and also lost it. The picture both the losers give of the victor and his methods is identical.

The German was no ordinary pilot. He was Ernst Udet, who lived to score sixty-two victories in all and play a significant part in the formation of Hitler's Luftwaffe. On that particular dawn he was stalking the enemy's observation balloons when with war-wise eyes he picked up a small dot coming out of the west. Just one machine, a loner like himself, looking for a fight, like himself. Armed with fixed, front-firing guns, the pilot who let the enemy get behind him was lost. The two machines passed each other at exactly the same height, a hair's breadth apart, and swept round in steep turns in an attempt for the other's back. Neither succeeded, and again they passed so close that Udet could read one word written in black letters on the brown fuselage of the French machine – *Vieux* . . . So this was Guynemer, victor then in thirty or more combats, who had the words *Vieux Charles* ('old Charlie') painted on his plane. Udet was close enough to make out the man's pale face under the leather helmet.

'Guynemer,' wrote Udet:

> Guynemer, who always hunts alone, like all dangerous predators, who swoops out of the sun, downs his opponents in seconds, and disappears. Thus he got Puz away from me. I know it will be a fight where life and death hang in the balance . . . I do a half loop to come down on him from above. He also starts a loop. I try a turn, and Guynemer follows me. When I come out of the turn, he gets me in his sights for a moment. A metallic hail rattles through my right wing and rings out as it strikes the struts. I try everything I can, the tightest banks, turns and sideslips, but with lightning speed he anticipates and reacts at once. Slowly I realise his superiority. His aircraft is better, he can do more than I, but I continue to fight. Another tight turn. For a moment he slides into my sights. I push the gun button . . . but the gun is silent. A stoppage!

Flying the plane with his left hand, Udet tried with his right hand to pull the cartridge belt and so get the next round into the breach. But it was immovable. To dive away would be suicide. He did his best to avoid giving Guynemer a target, turning all the time, as tightly as possible.

> I never had such a tactically agile opponent. For eight minutes we circle around each other. The longest eight minutes of my life. Then, inverted, he flies above me. For a moment I let go of the stick and hammer at the ammunition feed mechanism with both fists. Primitive, but sometimes it works. Guynemer must have seen me do this, must know I'm helpless. Again he skims over me, nearly inverted. Then it happens: he sticks out his hand and waves to me, waves gaily, and dives away towards his own lines. I fly home. Numb.

When Udet told this story, some cynics said: 'Oh, Guynemer had a stoppage himself then, that's why he didn't kill you.' Others claimed the Frenchman did it from fear – he thought Udet, in despair, might ram him. Udet himself declared: 'I don't believe any of them. I believe to this day that a bit of chivalry from the past has continued to survive. For this reason I lay this belated wreath upon Guynemer's unknown grave.'

The German ace's verdict agrees with Guynemer's own stated views on the ethics of shooting down unarmed or badly armed two-seaters: 'It is simply murder for the fast chaser aeroplanes to bring down the poor old observation planes; but in view of the consequences of the observation to artillery and infantry it is necessary to repress one's natural repugnance to engaging in such unequal combats and attack the slower planes with all one's strength.' On military grounds he should not have spared Udet, but very likely he did.

Guynemer's encounter with the British pilot came in the last week of life the Frenchman had. A French Wing was operating in a British sector and the British sent them a complete collection of their current aircraft, so that there should be no unfortunate mistakes of non-recognition. Cecil Lewis, MC, a pilot in the crack No. 56 Squadron, the 'Anti-Richthofen Circus', was to demonstrate the best aircraft Farnborough had produced, the SE5 scout. This is his account:

> The crack squadron in the French Wing was the 'Storks'. Its leader was the famous French ace, Guynemer. He was a slight, pale, consumptive-looking boy with black curly hair and a timid manner. He had three machines: a standard Spad, a high-compression Spad, and a larger special machine made by the same firm with a 200-h.p. Hispano engine (similar to the one

fitted to the SE). But this machine was unique in having a four-pound Pom-Pom firing through the hollow propeller boss.

A race was held between the two special Spads and the SE5. Their speeds were almost identical, but the high-compression Spad climbed quicker. After the race was over, Guynemer and I held a demonstration combat over the aerodrome. Again I was badly worsted. Guynemer was all over me. In his hands the Spad was a marvel of flexibility. In the first minute I should have been shot down a dozen times. Nothing I could do would shift that grim-looking French scout off my tail. Guynemer sat there, at about thirty yards' range, perfectly master of the situation. (In self-justification, I feel I must add that both the Sopwith Dolphin and the Spad were more manoeuvrable than the SE5. So that, given equal flying ability, they would win. Given still greater skill, the SE5 was right out of it). At last, we came down, landed, shook hands . . . Only a week later Guynemer was shot down and killed.

The last week of life for Guynemer was one of prolonged crisis. His closest friend, Captain A. Heurteux, was badly wounded. His favourite Spad was unserviceable. He flew other Spads and on one single day he flew three of them and force-landed each one, either with engine failure or a structural failure. Twice his guns jammed in battle. He flew five long patrols in four days, and failed to score a victory: washouts, all of them. From worry, he could not sleep; he paced the floor of his room all night. He suspected that he was being whispered about in the mess, that his comrades thought his nerve had gone. In fact, they were so concerned that they telephoned the Air Ministry in Paris, saying that Guynemer was ill and should be banned from flying for his own good.

The Air Ministry acted with speed and diplomacy. They sent two investigating officers, Captain Felix Brocard, former CO of the 'Storks', and Commandant Jean du Peuty, head of the Aviation Staff, GHQ. They arrived at 9 am on 11 September 1917. Guynemer was not there to meet them. He had taken off at 8.35 am with Pilot Officer Benjamin Bozon-Verduraz. Only Bozon-Verduraz came back.

His story was not very helpful. It had been a morning of low cloud and poor visibility, brightening slightly later on. In German air space near Poelcappelle, Guynemer saw a DFW two-seater, and the two Spads carried out a combined attack to distract the rear gunner. He fired back, the pilot spun the big machine, but as he recovered Bozon-Verduraz attacked in his turn. He, like Guynemer, failed to hit. Then Guynemer came in again, just as his companion spotted eight German aircraft above – possibly a trap. He broke off from the two-seater and headed towards them, but they carried on without apparently sighting either of the two Spads. Bozon-Verduraz was now unable to see either Guynemer or the DFW. There was no sign of them in the air, no crashed plane on the ground below. He landed at 10.25 with his negative information.

For ten days there was no word of Guynemer. Then Allied newspapers began publishing the fact of his disappearance. But it was not until a month after his last flight, on 11 October, that a German local newspaper covering the Cologne area reported the death of Leutnant Kurt Wissemann, the conqueror of Guynemer. The paper published extracts from Wissemann's last letter home, in which he claimed to have shot down the famous French ace. Ironically, he had added, to reassure his parents: 'Do not be anxious, as I can never have a more dangerous enemy.'

This reticence on the part of German officialdom was curiously uncharacteristic. The British Air Board's quick and unconfirmed claim for the death of Immelmann was much the more usual thing. Wissemann's killers, for instance, were known: Captain G.H. ('Beery') Bowman and Lieutenant R.T.G. Hoidge of the crack No. 56 Squadron, the unit to which Cecil Lewis also belonged.

At length, through neutral (Spanish) sources, the German authorities were pressed for more information, which proved contradictory and inaccurate in a minor detail – the date. The final story they released gave the correct date, 11 September (not the 10th). Guynemer's death wound was in front: a bullet had ricocheted off the gunsight and struck him in the forehead; another bullet had removed the forefinger of his left hand. 'The wounds appear to be consistent with the dead pilot having flown into the return fire from the rear gun of a two-seater. The German airmen express their regret at having been unable to render the last honours to a valiant enemy.' They had earlier said that Guynemer had been buried with full military honours in the cemetery at Poelcappelle. Wissemann, who could not now be questioned, gave the site as near St Julien, south of Langemarck. The grave never was found, although the area was captured by the British Army shortly after; perhaps that was the reason – the offensive had steamrollered both crash site and grave into anonymity.

Before W.E. Johns began writing schoolboy versions of 'Biggles', he edited an aviation magazine (and accepted my own first short stories for it) and also wrote a book called *Fighting Planes and Aces*. He had himself flown DH4 day bombers in the Great War, so he knew what he was talking about. He found the German treatment of the missing Guynemer most curious because, he said, the Germans made a point of identifying the aircraft and airmen who came down behind their lines. Indeed they did, and the victorious pilot often landed beside his victim in order to confirm it, or at the very least drove there by car as soon as he landed. How was it that they did not do so in this case?

A tentative answer may be suggested: that the aircraft crashed in an area inaccessible to motor traffic, and that nobody bothered much about identification until pressure was put on the German authorities via neutral sources, once the Wissemann story had leaked out in the local press. Quite simply, German officialdom may not have been absolutely convinced that Wissemann had indeed shot down Guynemer, and did not wish to put out a story that might be easily discredited.

One thing is certain, however, and that is the reason for the death of Guynemer: battle exhaustion. He was worn out and undergoing the recklessly suicidal stage of that experience. If he had delayed his take-off by only twenty-five minutes, Brocard and du Peuty might have been able to persuade him to take up the instructional duties which would have saved him for France.

On the British side, Albert Ball was the contemporary of Guynemer, flying at the same time and in the same lone-wolf, aggressive style. Initially, he was just as bad a flier. But he was younger still, born in 1896, so he was twenty years old at the height of his career. In other respects, however, he closely resembled Immelmann. Born at Nottingham, the son of a business man, he went to a minor public school and developed technical hobbies – engineering, photography, radio. His father taught him revolver shooting while he was still very young, and because for air fighting the skills of snap-shooting at fast-moving, jinking targets are both more difficult and more essential than airmanship, this training was probably a key factor in Ball's career. At eighteen, he had started, and was running, a small engineering and electrical business. Then, for him, as for millions of others, August 1914 altered everything.

Like Immelmann and Guynemer, he started in the ranks, quickly became a sergeant and was then commissioned. Like the Saxon boy, he had reservations about the army, writing in 1915: 'It is rotten to attend this Court Martial, for the poor men get delt [sic] with in a rotten way, by the officers and I do not think that the old fools know what they are doing, for they are making a perfect bilge of it and are having to look up every little point.' Ball's spelling was very individual, too. His ambition was now to join the Royal Flying Corps, so he paid for a course of civil flying lessons at Hendon and by this means managed to get in. His first instructors were not awed by his promise. One of them told him that he should look for a good flying school for girls, and join it at once, as he would never let him fly again.

By March 1916, he was in France and flying a BE2c on artillery spotting duties. These aircraft were pure 'Fokker fodder', being inherently stable (an excellent characteristic for observation and photography but not for dog-fighting), and were being shot down in droves for the next year or so, both by the Fokkers and the German 'Archie' (anti-aircraft fire), which was unpleasantly accurate. After seven weeks of this, Ball was beginning to mention 'nerves' in his letters home. At the same time, he tended to go off, without orders, after any German aircraft he saw, and one observer testified that Ball saw them long before anyone else, when they were just a speck in the sky. More than that, he seemed to sense when enemy aircraft were about. He passed it off as a joke, saying: 'I suppose I smell them.' This instinct and the uncannily accurate long sight must have been further key factors in his career.

There was then no such thing as a fighting squadron, or even a fighter aircraft. Attached to squadrons, however, were some faster single-seaters which were called 'scouts', because the main object of the air force was reconnaissance for the army. Sometimes they were called 'fighting scouts', a concept derived from the role of cavalry as a reconnaissance screen; although it was not desirable for scouts to get engaged, they had to be capable of fighting their way out of trouble if they bumped into it. The concept of building aircraft specifically for fighting, then massing them together and trying to drive the enemy out of the air, so that your air operations continue but his do not, was still only an idea in the minds of a few young men. Ball was a young man, who, by borrowing a fast 'scout' whenever he could, and using it recklessly to destroy or force down enemy aircraft, helped to create the concept of the fighter aircraft and the fighter pilot; and when he was killed, in 1917, it was as a leader in a real fighter squadron. His score was not so high as those of 'aces' who came later, but it was harder earned, because he was a pioneer.

One important factor which perhaps affected the mystery surrounding his death was publicity. The French started the 'ace' idea; the Germans refined it and graded it; the English ignored it, until the press in Nottingham, by featuring many 'local boy' stories about Ball, set him on his way to becoming a national figure, in effect a British 'champion' to rival Guynemer and defy the German heroes. RFC officialdom felt, perhaps rightly, that to concentrate publicity on only a few airmen, all of them pilots, and all of them single-seater pilots moreover, would not improve the morale of the crews of the reconnaissance aircraft, who might feel themselves slighted. However, the public wanted invincible champions.

The odds, of course, were totally against it. On his last leave in 1917, Ball told his father: 'No scout pilot who does any serious fighting, and sticks it for any time, can get through.' Earlier in his career, in his letters home, he had made no secret of the strain. 'I shall be glad when it is all over, for at times I feel very wonky and run down. But my pecker is always up, and if my health only holds on tight I shall be O.K.,' he wrote to his mother. To his father: 'At night I was feeling pretty rotten, and my nerves were feeling quite poo-poo.' He felt so bad then, that he asked his CO for a short rest – and, for punishment, was sent back on BE2cs. Later, when the RFC had realised his worth to them, General Trenchard himself would ask Ball if he wanted fourteen days' leave. Ball's own antidote was sleep, and plenty of it. Later he had a small hut and garden on the aerodrome, to which he would retire to be quiet and practise playing the violin.

What Ball was doing was deadly dangerous. He had begun to specialise in attacking German formations and picking off one machine before the others could do anything; he would describe these as 'ripping' fights, which he had really enjoyed, except for the necessity of killing people. He was particularly saddened when he sent an enemy down in flames, because unless the pilot jumped, he would burn to death, there being no parachutes yet. He had seen it happen, and he knew it could, and almost certainly would, happen to him.

Ball was sent home to instruct during the winter of 1916/17. In the spring he joined No. 56 Squadron, which was to be led by experienced scout pilots and equipped with the new SE5s. Ball had himself designed a fighter aircraft, which was being built at this time, the Austin-Ball scout. The SE still had teething troubles, but the Squadron had to go to France, because of the beating the RFC's obsolete types had taken during 'Bloody April' from the new German fighters, particularly the group led by Baron Manfred von Richthofen, which had now acquired the nickname of the Richthofen 'Circus'. No. 56 did reasonably well until 7 May, a sunny day which soon became overcast and stormy, with thick layers of cumulus cloud between 2000 and 8000 feet. At 5.30 pm, in spite of worsening weather, eleven SE5s were ordered off into a darkening sky swarming with German fighters, including those of Richthofen. Two hours later, two SE5s struggled back, and half an hour later, three more. And that was all. None of the three flight commanders, of whom Ball was one, had returned.

Some of the others were down safe, or down wounded, but three never came back, and among those three was Captain Albert Ball. Probably the last British pilot to see him was Cecil Lewis (who was to have his mock duel with Guynemer the same year). Ball's machine was distinctive, with a red spinner to the propeller and flight commander's streamers. 56 Squadron, stacked in layers, got involved with several German units, including Richthofen's. They became scattered, but were joined by other British fighters, including Spads and some Naval triplanes. On both sides reinforcements were sucked in from above, and so the fight was even more confused and impossible to narrate rationally than was normal. In his book *Sagittarius Rising*, Lewis gives a vivid impression:

> It would be impossible to describe the action of such a battle. A pilot, in the second between his own engagements, might see a Hun diving vertically, an SE5 on his tail, on the tail of the SE another Hun, and above him another British scout. These four, plunging headlong at two hundred miles an hour, guns crackling, tracers streaming, suddenly break up. The lowest

> Hun plunges flaming to his death, if death has not taken him already. His victor seems to stagger, suddenly pulls out in a great leap, as a trout leaps on the end of a line, and then, turning over on his belly, swoops and spins in a dizzy falling spiral with the earth to end it. The third German zooms veering, and the last of that meteoric quartet follows bursting . . . But such a glimpse, lasting perhaps ten seconds, is broken by the sharp rattle of another attack . . .

The clouds, however, were a different matter. They could be described: 'The May evening is heavy with threatening masses of cumulus cloud, majestic skyscapes, solid-looking as snow mountains, fraught with caves and valleys, rifts and ravines – strange and secret pathways in the chartless continents of the sky.' Among these clouds the scattered fighting drove lower, as the outnumbered and outfought British scouts tried to escape. Ball, typically, when last seen, was not trying to get away. 'I believe I was the last to see him in his red-nosed SE going east at eight thousand feet,' wrote Cecil Lewis. 'He flew straight into the white face of an enormous cloud. I followed. But when I came out on the other side, he was nowhere to be seen.'

Captain A.M. Crowe also thought that he might have been the last British pilot to see Ball, at about the same time. Noticing Ball fire two red warning Very lights, he looked round for enemy, and saw only a Spad and, far off, a triplane. Then, following Ball, he saw him nose over into a dive, and at last saw the German single-seater which Ball was after. They disappeared eastward, fighting, into 'a very heavy bank of cloud' which Crowe entered also; like Lewis, when he emerged the other side, there was no sign of Ball's SE5 or the German.

Ball was a quiet man. Cecil Lewis wrote: 'He was not a stunt pilot but flew very safely and accurately . . . He never boasted or criticised, but his example was tremendous.' The example, however, was to fly to the limit when in action. Sergeant-Major A.A. Nicod described his returns when he was with No. 11 Squadron.

> We would gaze with astonishment at his machine after a scrap – no ammunition, petrol tanks empty – marvelling how he managed to get back. Fabric ripped, struts and longerons shattered, wires broken, bullet holes all over the machine. Yet

he would give orders for his tanks to be refilled and ammunition to be replenished, and even jocularly suggest to his rigger to tie his struts together.

When, this time, Ball failed to come back at all, there was the ominous possibility that he had run out of ammunition while still fighting; he had risked this many times before. Or he might have left insufficient fuel to return and now be a prisoner. But the Germans did not at first claim him and 56 Squadron dropped notes on Richthofen's aerodrome at Douai, asking for news.

Later, the Germans said that Manfred's brother, Lothar von Richthofen, had shot him down in single combat, and that Ball had smashed the German ace's fuel tanks, so that the red Albatros had come down, too. They dropped a brief message, first: 'Captain Ball was brought down in an air fight on May 7th by a pilot who was of the same order as himself. He was buried at Annoeullin.' In Manfred's autobiography the senior Richthofen gave a detailed account, beginning: 'Captain Ball flew a triplane and encountered my brother flying by himself at the Front. Each tried to catch the other. Neither gave his opponent a chance. Every encounter was a short one . . .' Eventually, the two fighters flew head on at each other, and the German's petrol tanks were holed, so he switched off his engine quickly and went down for a dead-stick landing. Lothar looked around for his enemy, then spotted it below him. 'He saw the triplane falling down in a series of somersaults. It fell, fell, fell until it came to the ground, where it was smashed to pieces . . . Captain Ball had been shot through the head. He carried with him some photographs and cuttings from the newspapers of his town, where he had been greatly feted.' That was the German story published during the war.

The confidential reports told the same tale more briefly. Lothar had reported: 'On May 7th I had a combat with many triplanes. One of them attacked me in a very determined manner. We fired a great deal at each other, and during the combat he came very close. He came down under my fire. My machine was damaged, and I landed with a dead prop, near the hostile machine.' The German Air Service did not grant victories on the pilot's word alone. There had to be three witnesses, and three witnesses Lothar duly produced. Leutnant Hepner, of kite-balloon Abteilung I, testified that he saw a triplane fall out of control. Leutnant Hailer of *Aieger-*

Abteilung 292 confirmed the victory. *Flakgruppen* 22 provided another witness to seeing a triplane shot down. A fourth witness could say only that he saw an English machine crash at a farm near Annoeullin and a German machine force-land about a mile and a half away. This last testimony made it seem unlikely that Lothar checked the crash in person.

Cecil Lewis recalled that a flight of Sopwith Triplanes from No. 8 Squadron, of the Royal Naval Air Service came down to help the SE5s. They were feared by Albatros pilots because of their manoeuvrability and rate of climb, and had obtained a reputation similar to that to be earned by the Fokker Triplane later. There is no reason to doubt that Lothar von Richthofen had a hard fight with the pilot of a Naval triplane. But not with Ball, because he was flying an SE5 which, although a new type to the Germans, was quite definitely a biplane and not to be mistaken for a triplane (unless what one was looking at was a mass of shattered wreckage on the ground).

Thus began the Albert Ball mystery, not easy to resolve, for the fight had been unusually tempestuous and scattered, both in height and space, and among towering cloud masses, with many different squadrons engaged on either side. Unlike Guynemer, however, Ball did not disappear. I seem to recall seeing some of his effects, including a crushed cigarette-case, in the Imperial War Museum in the 1930s. In 1933, R.H. Kiernan published a biography in which he quoted some evidence, not first-hand, that Ball may have been brought down by AA fire. After publication, he received a letter from an RFC officer who had been shot down three days before Ball and was being held prisoner at Douai:

> After I had been in solitary confinement for about a week or less the German intelligence officer came to see me, a Bavarian, quite a decent fellow, I have forgotten his name. He asked me whether I knew Captain Ball, as someone's body had been found whom they thought from his decorations might have been Ball. I said yes, as I should have said in any case as I wanted to get a spell outside. But it happened that I had met Ball briefly at Nottingham when he was on leave. He was pretty badly smashed up, but I had no difficulty recognising him. This German officer told me that no one had reported shooting down an English machine at that time on that sector of the front, and

> the conclusion was come to that he had been hit by anti-aircraft guns. He also told me that he was to be buried with military honours.

This letter was printed in *Popular Flying*, edited by W.E. Johns, to which I then contributed occasionally.

Many years later, in 1961, Frederick Oughton published *The Aces*, a book in which he was able to quote at length one of Lothar's witnesses, the former Leutnant D.R. Hailer, now a civilian. Herr Hailer now said that he saw an SE5 (i.e., *not* a triplane) flying upside-down through the clouds but apparently under control until a German battery opened up and it crashed. He went to the site (there had been no fire) and found that the body, with some limbs broken, was that of Captain Ball, according to his wallet and identity disc. He could find no evidence that either the pilot or the aircraft had been hit by German fire (although he met an artilleryman who claimed to have shot it down). Later, he said that he and another officer rearranged the evidence, by firing bullets into the wreckage, so that it would appear that the aeroplane had been brought down by machinegun fire.

Oughton also contacted Gerald Maxwell, who had been a pilot in Ball's flight of No. 56 Squadron and scored twenty-seven victories. With the advantage of an RAF career which covered the Second World War also, Maxwell could now point out how ignorant they all had been in that first war, of the potential danger of some types of cloud formation; most had regarded cloud simply as cover, in which to ambush or to escape. He thought that there was an element of propaganda in the German claim for Lothar von Richthofen and believed that the wings of Ball's SE5 could have iced up in cloud and that he had fallen out of the cloud, upside-down and too low to recover, more or less as described by Leutnant Hailer.

Another not entirely dissimilar reason could have been fatal structural damage to the SE caused by turbulence inside the cloud. The development of gliding and soaring between the two wars produced evidence for tremendously powerful forces being at work within certain types of cloud formation. It is also true that the actual machine he was using on that day, according to Kiernan, was the repaired SE on which he had gained his thirteenth, fourteenth and fifteenth victories. The repairs could have been structural weak points contributing to a collapse.

As for Lothar von Richthofen, he claimed to have fought a triplane which was very well handled, and we may believe him. But then he lost sight of it, after they had charged each other head-on, firing and then breaking away violently, partly inverted. After that, he had to search the sky, anxiously, for his opponent; and at length saw an aeroplane, which he presumed to have been the one he had been scrapping with, falling earthwards out of control. It may have been too far away for him to be certain that it was a triplane, let alone the one he had been fighting. *That* machine could have been Ball's SE5. We shall never know.

Rittmeister Manfred Freiherr von Richthofen was the top-scoring fighter pilot of the war and the most controversial. Born in 1892, he was twenty-two years old in 1914. His family were landed gentry from East Prussia, whose spare-time country pursuits were riding and shooting. Manfred's father was one of the few members of the two families who became a professional soldier, however, and it was he who decided that young Manfred would join the Cadet Corps when he was eleven years old. 'My wishes were not consulted,' wrote Manfred. 'I found it difficult to bear the strict discipline and to keep order. I did not care very much for the instruction I received. I never was good at learning things . . . So I worked as little as possible. On the other hand, I was pretty fond of sport.'

On the outbreak of war he was an Uhlan officer on the Russian front. He gave a very modest account of his services there, although he says he enjoyed it, and then told with the utmost frankness the story of how he led his cavalry troop straight into a French ambush soon after being transferred to the west. The completely objective way in which he described the appalling and deadly muddle into which he got his men is a clue to Manfred's character.

In the spring of 1915 he transferred to the Air Service as an observer, first in the east then in the west. He had his first air combat in September 1915, engaging an enemy aircraft with a rifle. At Christmas he passed his tests as a pilot (although he crashed the aircraft on his first solo) and was posted to a two-seater unit, where he scored an unconfirmed air victory. By September 1916 he was flying an Albatros D II single-seater in the fighter squadron formed by Oswald Boelcke, Jagdstaffel 2.

By June 1917, the principle of concentration had produced the grouping of a number of squadrons into a super unit, the *Jagdgeschwader*. The first of these to be formed, JG 1, was led by Richthofen, whose orders were 'to attain air supremacy in sectors of the front as directed'. The air organisation was tied into an observation and reporting system on the ground, so that the fighters could be directed on to their enemies rather than fly wasteful standing patrols. The air units were moved from sector to sector, as the military situation demanded, often obtaining a local superiority over the enemy although in total numbers the German Air Force was much outnumbered by the Allies. The prime interception targets for the German fighters had to be the Allied photographic and artillery observation aircraft which were directly affecting the land battle. The fact that many of these were ill-armed aircraft unsuitable for war was hardly Richthofen's fault: nor was the fact that, in order to do their job, they had to fly over the German side of the lines. Yet Richthofen's detractors disparage both his courage and his skill on these grounds.

One of the most biased was the fiery Welshman, Ira Jones, himself an ace, once he had got away from flying in BE2cs, who wrote that many British pilots were superior to Richthofen (probably including himself among that number). He seems to have regarded air fighting as a kind of personal duel, with everything fair and equal, and no consideration whatever for military objectives. He attacked Richthofen for letting the enemy come to him (just as the RAF were to do in the Battle of Britain). Jones wanted the Germans to venture more over the British lines, where any German machine and pilot which force-landed would be irrevocably lost, and thought they were very unsporting, Richthofen in particular, because they failed to run the war his way. He went further, and wrote: 'Richthofen also took care that he did not fight a machine of the calibre of his own, unlike the redoubtable German Jew, Werner Voss, who literally tackled anything under the sun. The only time Richthofen appears to have come any great distance over our lines was the day he met his doom.' Although deeply biased, Jones was thoroughly honest, and added: 'It is my belief that the fatal shot was from a rifle.' The Welshman was risking a charge of heresy, for the Royal Air Force claimed the credit.

Jones was less vicious than Mannock, the Irishman, the top-scoring British ace with seventy-three victories, who, just like Richthofen, did not believe that it was consistent with his duty to give an opponent an even break. On hearing of Richthofen's death, he said: 'I hope he roasted the whole way down.'

A more informed view was that of the Dutchman, Anthony Fokker, who knew many of the German aces well and was old enough to make mature judgements. He did not blame Richthofen for being 'imbued with the usual ideas of a young nobleman', for that was a matter of birth. He praised him for the 'superior will' which made him in the end a good pilot, for he had no instinctive feel for flying.

> Ultimately, Richthofen became an excellent flyer and a fine shot, having always done a lot of big game hunting. But whereas many pilots flew with a kind of innocent courage which had its special kind of magnificence, Richthofen flew with his brains, and made his ability serve him. Analysing every problem of aerial combat, he reduced chance to the minimum. In the beginning his victories were easy. Picking out an observation plane, he dived on it from the unprotected rear, opened up with a burst and completed the job almost before the enemy pilots were aware of trouble. It was something of this machine-like perfection which accounts for his near death in 1917 after his fifty-seventh victory.

The fight was against six FE2ds of 20 Squadron, but 'pushers' with the engine behind the crew and a gunner in the front cockpit handling a forward-firing free gun (unlike the poor BE man, trapped in a cage of wings and wires). As Richthofen led his Albatros DVs in to the attack, a British gunner opened fire at the impossibly long range of 300 yards at the leading German machine – and hit it. One bullet struck Richthofen's skull, exposing the bone, and making him blind. As the machine fell, he switched off the engine and threw away his goggles, but still the Albatros fell and still he could not see. Gradually sight came back and he managed to land. A spell in hospital at Courtrai followed, but he had dizzy spells for some time afterwards, and his psychological armour had been broken. Now Richthofen knew that he too was mortal. 'Manfred was changed after he received his wounds,' his mother is reported to have said.

But as a *Geschwader* leader, he was still superb. Anthony Fokker noted that although 'Richthofen knew little or nothing about the technical details of aeroplanes . . . and was not even interested', he would hold a conference immediately after each air battle in order to discuss the tactical lessons. Pilots could incur his censure in one of two ways, by being either too aggressive or too ready to pull out of action. But his comments were perfectly objective, delivered in a calm, good-humoured manner, with no foul language. What heartened his men, as he spoke, was the realisation, as one of them said, 'that he never lost sight of us even when fighting for his life . . . We knew we could count on him like a rock.' But he also had a touch of ruthlessness, as Ernst Udet noted: 'His estimate of a man was formed by what that man achieved for the cause, and whether he happened to be a good fellow or not was of a secondary consideration . . . If you were a failure he dropped you without a second's hesitation, without a flicker of an eyelid.' Unlike many with less responsibilities, he did not let off steam in the evenings in the mess, but usually retired to his room to think about the next day's flying.

His last day came during the German spring offensive in 1918, the last throw of Imperial Germany to win the war before American intervention became overwhelming. JG 1 flew fighter cover over the advance. The opposition was much better equipped than it had been in the spring of 1917. Richthofen's victims were no longer the old BE2cs, but the best fighters in the RFC – the Camel, the SE5A, the latest type of Spad, the Bristol Fighter. Richthofen was now flying a Fokker Triplane, not a fast machine, but light, low-powered and agile, a joy to handle after the heavier Albatros DV. On 21 April, he was ordered up to intercept two RE8 photographic machines of No. 3 Australian Squadron. He took off with five companions, but was later joined by part of another squadron from JG 1. Some of their machines were Albatros DVs. Richthofen intercepted the two RE8s, which were busy photographing the German positions, and led three other triplanes down to the attack, while the remainder stayed above as top cover. From that moment on, someone or other was claiming the credit for having shot him down, all in good faith, and several different people were decorated for it. Most of them did not know at the time who they were shooting at, but only that the pilot flew a red triplane. That was not conclusive, because many of the Baron's companions flew red Fokker triplanes, each with a small splash of another colour to differentiate it.

E.C. Banks, the observer in one of the RE8s, testified that the photographic machines were not caught unawares, but skilfully protected each other for six to eight minutes, shooting down one red triplane and damaging another. Banks was convinced that he had got Richthofen. Certainly the Australians drove him off, for the Baron's next enemy was a formation of eight Sopwith Camels from 209 Squadron, led by a Canadian, Captain A. Roy Brown, who had spotted the telltale bursts of British AA fire which the German aircraft were attracting.

Brown's squadron had already engaged two German reconnaissance machines and shot one down; now they went for the 'Circus'. One man on either side was a comparative novice, ordered to stay out of the fighting. The German was Wolfram von Richthofen, a cousin of the Baron's. The Canadian was Lieutenant Wilfred May, who stayed above the battle until fatally tempted to attack a triplane gaining height, which may have been flown by Wolfram. May's attack went wrong, he missed the triplane and found himself in the middle of a vicious, close-quarter merry-go-round. From excitement, instead of firing the regulation brief bursts, he kept his finger on the trigger and irrevocably jammed both Vickers guns. Then May found a red triplane behind him, and dived the Camel headlong for the ground and the British lines (held in this sector by the Australians). Brown saw this, and came to the novice's rescue. His combat report read: 'I dived on a pure red triplane which was firing on Lieutenant May. I got a long burst into him and he went down vertical and was observed to crash by Lieutenant May and Lieutenant Mellersh.' May, equally brief, reported: 'I was attacked by a red triplane which chased me over the lines, low to the ground. While the triplane was on my tail, Captain A.R. Brown attacked and shot it down. I observed it crash into the ground near Vaux-sur-Somme.' As a result of these reports, an RAF communiqué announced: 'Captain Manfred von Richthofen, who is credited by the enemy as having brought down 80 Allied machines, was shot down and killed behind our lines near Corbie by Captain A.R. Brown, 209 Squadron.' It all seemed absolutely conclusive.

A German balloon observer witnessed the incident, not knowing who was in the plane, and reported: 'Red triplane landed on hill near Corbie. Landed all right. Passenger has not left plane.' This simply did not fit with the behaviour of Brown's triplane, which 'went down vertical'. They could hardly be the same machine, and doubtless they were not. Neither Brown nor the German observer claimed that the victim was Richthofen, but merely that the machine was a red triplane. In their reports, the crews of the two RE8s also claimed just a red triplane. It was only when news came through that Richthofen actually had been shot down behind the Australian lines that excitement began to mount.

A great many Australians witnessed the undoubted end of Manfred von Richthofen, because the Baron chased May for several miles and only a few hundred feet up across their Sector, as the wretched novice with jammed guns fled for his life up the valley of the Somme. He was so close to the fleeing Camel that the Australian ground machineguns – the area was packed with them – dared not fire for fear of hitting the Canadian pilot. To begin with, the Baron was putting short bursts into the Camel. Then, first one and then the second of his guns jammed. Recklessly, he continued chasing May unarmed, in spite of the fact that he had got himself downwind of the German lines and would have to fly back against the wind. All-out, the Fokker triplane could just touch the 100 mph mark. For once in his life, he was allowing the odds against him to stack up one by one to form a deadly combination.

Brown was no longer in sight. After he had fired one burst at the triplane, and seen (he thought) the pilot slump down, May and his pursuer had vanished from his view behind some trees, and as Brown himself was in a bad way from stomach trouble and dysentery, and his machine low on fuel and ammunition, he carried on in a direct line for the home aerodrome at Bertangles. This explains why the Australian ground forces saw only the two aircraft – May's desperately jinking Camel and the pursuing triplane skilfully jockeying every move behind him, and at the last with silent guns. But May, without knowing he had the Baron behind him, had almost given up. He took the Camel round a bend in the river Somme, so close that a slight touch on the controls would have taken him into the water, and was tempted to do just that and so end it all, when he realised that Richthofen was no longer behind, but had turned and roared over the hill to catch him as the Canadian, still following the river, came round the other side.

Or so May recalled. What seems certain is that, unable to clear his jammed guns, and perhaps having been hit by fire from the ground, Richthofen broke off the chase by turning right over

Corbie Hill, the western extremity of Morlancourt Ridge. Up to now he had been as close from May's tail as thirty feet, but as the triplane broke away the ground gunners had a clear shot at him. Everything opened up, from Lee Enfield rifles to Lewis and Vickers machineguns on high-angle mountings. The 24th MG Company started firing, so did the machineguns of the 44th Battalion, and then the Lewis gunners of the 53rd Field Battery. The triplane was going round in a tight turn to the east, wings pointing almost directly at the ground, when it was seen to stagger and something fell away from it – Richthofen's goggles, which he had torn off, as he had before when he had been wounded.

The triplane came down to the ground in a long slant, wings level now. It struck, bounced ten feet back into the air as if it had just made a bad landing, then crashed into a pile of mangels in a field which cushioned the shock. The machine was not too badly damaged, and as the German balloon observer reported, the 'passenger' did not get out.

Within a minute, curious Aussies were swarming round the red machine, anxious for souvenirs, although the actual crash site was in view of the Germans. A little cloud of dust hung over it. In a short time, the machine had been stripped of gadgets and fabric, thus removing all evidence of how many bullets had actually hit the triplane. The pilot was clearly dead, much cut up about the face from injuries received in the crash (a few witnesses wrongly thought he had been shot in the face), but he had been shot once only, and fatally, by a bullet entering low down on his right side at the back and exiting high up on the left. The damage was so great that he could not have flown any distance after receiving it.

Lieutenant George M. Travers, MC, of the 52nd Australian Battalion, saw the triplane go down after hearing a burst of ground machinegun fire and said that he was in the first three to reach the wreck, the others being Captain Cruickshank and a runner. Cruickshank, a staff captain of 11th Infantry Brigade, took the pilot's watch and papers to Brigade HQ, and they identified him as Manfred von Richthofen. Travers kept a copy of his report, which was made public by accident many years later, long after Captain Brown had been credited with Richthofen's death.

Another officer witness who arrived early was Captain Donald L. Fraser, the Brigade Intelligence Officer. He, like Travers, had seen a red triplane chasing a Camel, with no other aircraft near, and also heard a burst of machinegun fire from the ground before the triplane crashed about 200 yards from him. Six men only had reached the wreckage ahead of Fraser, who had the pilot's body pulled clear, and searched. The Baron's identity disc told him who he was, but to announce it there and then would only worsen the souveniring which was going on.

Lieutenant R.B. O'Carroll of 44th Battalion was ordered to put a guard on the pilot's body. He saw a sergeant souveniring and said to him: 'Sergeant, you ought to be helping me, not showing a bad example.' He replied: 'Sir, I shot the bastard down so I ought to have something.'

Captain Fraser was ordered by his divisional commander, General Cannan, to identify the machinegunners who had brought down Richthofen, and he had no difficulty in locating the Vickers machinegun post manned by Sergeant Cedric Bassett Popkin of the 24th Australian MG Company. He also discovered two Lewis gunners of the Australian 53rd Battery who had fired on Richthofen 'from an acute angle'. They were Gunners Robert Buie and W.J. 'Snowy' Evans. All three had probably scored hits on the triplane, but who had fired the single shot that had killed the Baron? In Fraser's view, that man was likely to be Sergeant Popkin.

Weston, No. 2 of the Vickers, kept a diary which showed that Popkin had fired two bursts of eighty rounds. The first at 100–150 yards range had probably hit the triplane and alarmed the pilot, for Richthofen then banked sharply away towards his own lines; the second burst, from below and behind, coincided with the triplane going out of control and slanting 300 feet down towards the ground, where it crashed. This firing position was consistent with the single bullet wound Richthofen sustained. Neither Popkin nor Weston saw Roy Brown; they only noticed May and Richthofen.

The same was true of Gunners Buie and Evans. Buie saw the main dogfight and saw two planes shot down from it and a third corkscrew down but flatten out and make for the British lines. Then a German dived after it and both planes came very low; there were no other planes near them. The Australians' Lewis guns were mounted on posts, and being fed by a drum, there was no need of a No. 2, such as Popkin had, to feed in a belt. Evans had a clear shot first, then Buie, as May's Camel roared over. Both men had frontal shots at Richthofen as he approached and Buie saw his own bullets striking the nose and right side of the Fokker. He knew they were his own for certain, because every tenth round of the 45-round drum was a tracer. He could see those, like a red-hot cigarette-end, speed out of his gun muzzle and go arcing towards the nose of the red triplane. Richthofen himself was firing at May. Then his guns stopped, as he rocketed over the Lewis gun position, about fifty feet up.

The triplane appeared still to be under control and did not come down at once. It banked several times then lost height gradually until it hit the ground, not too violently, about 400 yards away. Major L.E. Beavis, commanding the battery, put in a claim on behalf of Gunner Buie, and Buie was congratulated on his success by Generals Rawlinson and Birdwood, which made it official. Popkin was less well served by his CO, who although he supported his sergeant's claim, put it in a week late. Considering the nature of Richthofen's wound, Popkin fired the fatal shot, but Buie probably hit the triplane. Captain Fraser, who investigated at the time on the ground, when it was plain where the gun positions were in relation to the wrecked aircraft, certainly came to these conclusions. There was no doubt concerning identity. Major Beavis had the body taken to his dugout, from where it was later removed by a vehicle from No. 3 Squadron. The Australian Flying Corps officer who came for the body did so on the grounds that Richthofen had been shot down by an RE8 of his own squadron. While he was removing the body and the wreck of the triplane (under German shellfire), Captain Roy Brown turned up, apparently trying to get confirmation from Major Beavis and his unit that he, Brown, had shot down Richthofen. He failed to get the confirmation that he sought – or had been sent to get by his superiors, for he was a modest man.

Richthofen's body was taken back to 3 Squadron's aerodrome at Poulainville and placed under guard. Most of those who mounted guard were keen souvenir hunters, some were keen amateur photographers, too. Consequently, the nature of the Baron's wounds was independently established even before the various posses of medical officers turned up to carry out the autopsies. The man who laid out the body found the bullet which killed him, next to a wallet; but did not report the fact. The first four doctors were all RAMC, but differed in ranks and allegiance. Colonels T. Sinclair and J.A. Dixon were from British Fourth Army, one a surgeon the other a physician. The juniors, Captain N.C. Graham and Lieutenant G.E. Downs, were attached to the RAF.

The surface wounds were explored with a probe, and the two Colonels found 'that there are only the entrance and exit wounds of one rifle bullet on the trunk.' The bullet had come out two inches higher than it had entered, having struck the spinal column. They gave no opinion as to how the shot had been fired, except that it was semi-frontal. The two junior officers attached to the RAF also found evidence of one bullet only. The exit wound was slightly higher than the entrance wound, but they did not think the bullet had been deflected by the spine (had they seen the bullet they might have changed their minds). They concluded that the shot had been fired from behind and slightly from the right of the fore-and-aft axis of the aircraft, and that it could not have been fired from the ground.

The matter did not rest there. Next morning two more doctors turned up. This pair were Australians, and they were accompanied by Major D.V.J. Blake and Captain E.G. Knox of No. 3 Australian Squadron, whose RE8s were claiming to have killed Richthofen. Colonel G.W. Barber was deputy director of Medical Services for the Australian Corps; with him was a senior colleague, Major C.L. Chapman. Barber found that: 'The bullet had passed completely through the heart and chest and, from its position, I formed the conclusion that it was fired from the ground and struck the airman as he was banking his machine, because the point of exit of the bullet was three inches higher than the point of entry.' Chapman reported: 'The bullet came out about an inch higher than it went in and might quite well have been shot from the ground.' A photograph of the exit wound shows it measuring two inches long by one inch wide, caused by the bullet turning end over end, perhaps (although exit wounds usually are much larger than entry wounds). All the weapons used – rifle, Lewis and Vickers – fired a .303 round, so without the actual bullet (and the matching of the grooves on it to a particular barrel) there could not be the kind of certainty demanded by murder cases.

There seemed to be the basis for an international incident, with Roy Brown the Canadian reluctantly pitted against the three rival Australian groups. There was even a solitary English claimant, another Lewis gunner, Sergeant A.G. Franklyn, of an AA battery attached to the Australians. Franklyn had indeed shot down a German aeroplane on 22 April, but he had misread the reports of

Richthofen's death and so believed that the Baron had been killed that day instead of, as he really was, on 21 April. No wonder Franklyn complained later that published reports of how Richthofen met his death bore no relation to what he himself had seen!

Who caused Richthofen's death can be answered with slight caution: almost certainly Sergeant Popkin, although Evans and Buie's shots may have caused him to swerve back into the Sergeant's line of fire.

What caused the death of Richthofen may be answered with greater certainty: battle exhaustion. As with Ball, as with Guynemer, so with Richthofen. The same symptoms are there. An irritating inability to knock down enemy aircraft – witness how Richthofen with four planes attacked the two RE8s, a very poor aircraft, and failed to get even one of them, possibly losing one of his own machines in the process. He then goes on to start a dogfight with Camels, picks a novice air-fighter like May, chases him right down to the deck and up and along the Australian front for miles – and fails to bring him down, even so. Richthofen, the master huntsman, the brilliant snap-shot! The pilot who kills, often enough, with a single burst from point-blank range! Here, he closes to within thirty feet of May's tail and still can't get a kill shot! Finally, the utterly uncharacteristic recklessness of it all – low down behind the enemy lines, packed with anti-aircraft weapons, and downwind of his own lines and safety. And on his own, instead of supported by his own pack. He had lived three years at the front – where many failed to last even three weeks – precisely because he calculated the odds accurately like a poker player, and fought to win.

Manfred von Richthofen was a very brave gentleman at the end of his tether and he was buried as such, glad though the British were to see his shadow lifted from the lines. They gave him a coffin, flowers, wreath, a Church of England burial service, a volley and 'The Last Post'. Some saluted. That night, the French broke into the cemetery, scattered the flowers, tore the wreaths, knocked over the cross, and partly dug up the grave.

When the Germans realised that Richthofen would never return, they opened what was in effect his last will and testament. It consisted of one line only: the name of his successor, Wilhelm Reinhard. Nothing for himself. Only his duty.

But the story did not end there. After the war, his family were not allowed to bury him at home, next to his father. Richthofen still belonged to the nation and must be buried in Berlin. In 1925, his remains were dug up from the French cemetery and taken to the German capital for a ceremonial funeral. After Hitler came to power, the Richthofen Circus was reborn, and this time it was to fight over England, during the Battle of Britain. Wolfram von Richthofen was to achieve high rank in the new Luftwaffe, as commander of the dive-bomber force. But now it was the Germans who had the power of numbers and were on the attack, the English who defended. In its turn, that war came to an end, but that was not the finish of the old Richthofen controversy.

When I was in my teens, all the British books gave Brown as the conqueror of Richthofen. The first notes of Australian dissent which I read appeared in the American magazine *Flying Aces* in January and December of 1935, both articles being by Arch Whitehouse (who had served with the RFC). In 'The Riddle of Richthofen's Death' and its follow-up 'Who Slew Baron von Richthofen? New Evidence Concerning the Greatest Mystery of the War', Whitehouse and his readers combined to present a convincing case for the Aussie ground gunners, which I was glad to draw on when, nearly thirty years later I came to write a book (*The Friendless Sky*, Souvenir Press, 1962) dealing with the development of military aviation up to 1918.

Meanwhile an American enthusiast, P.J. Carisella, had devoted most of his spare time for thirty years to the Richthofen story alone. He contacted scores of witnesses, and testimony piled upon testimony, eliminating Brown without further argument, for nobody from private to general saw any other Camel apart from May's. And Brown, of course, when his report is analysed, does not claim to have shot down Richthofen or even to have shot down a triplane – he claims merely to have fired at a triplane which was seen to crash by two other pilots, May and Mellersh. A German pilot sent to check on the Richthofen crash saw *two* planes down, not one, but was unable to get close enough to see detail. Carisella finds, with Captain Fraser, for Sergeant Popkin, assisted perhaps by Evans and Buie. This pleases the Germans, who like to think that Richthofen could never have been bested in an air fight, and displeases the RAF, for 209 Squadron still claim the Baron.

Understandably, Carisella developed a passion for collecting mementos of the crash, instruments, parts of the engine, bits of fabric 'souvenired' by Australians long ago. This led him to a most extraordinary discovery – the most unexpected find of all. In 1969 he was invited to a reunion of the surviving members of the old Richthofen Circus and on the way visited the sites in France where the Baron had been killed and where he was buried. He particularly wanted to recover the name plaque which had been attached to Richthofen's coffin. His information was, that when the body had been exhumed in 1925 for removal to Berlin, the coffin lid and plaque had been put back in the grave and reburied. It seemed pointless for such a unique relic to decay away in the earth, and he sought, and obtained, permission to dig. He had also been told, but had not believed, that the Germans had removed only Richthofen's skull, leaving the skeleton and coffin still in situ. That seemed impossible, in view of the importance attached by Germany to Richthofen.

But when the French workman had been digging for several hours in the hard chalk, the side edges of a coffin did in fact appear. Thinking that the plaque might have fallen into the empty box, they began to excavate inside it, and up came a spadeful of wet chalk – and an ankle bone! The workman shouted that it must be Richthofen, and handed the bone to Carisella, who was shocked and confused. He had started something he did not know how to finish. He took what in the circumstances seemed the logical decision and completely excavated the coffin, which proved to contain the lower part of a human body, with the skull missing. He handed over the remains to the German Embassy in Paris, who seem to have accepted them rather doubtfully, as might be expected. A strange story, but all it requires for acceptance is a belief that the two German workmen who exhumed the body in 1925 may either have skimped the job or been ordered to remove only the skull. The Richthofen mystery ends in another mystery.

Of all the aces who disappeared, initially without trace, quite the most unbelievable was Frank Luke. Like most, he was a mere boy, born in Phoenix, Arizona, in 1897 and dying in 1918, the same year he reached the front. His fighting career was very short. He never had time to grow up. Many pilots actively disliked discipline, particularly the army's idea of it. A simple patriotism seems to underlie the attitude of the aces, together with the normal instinctive desire 'not to let the side down'. There was perhaps some personal ambition also and, of course, no doubt about it, air fighting was exhilarating. At first, that is: until you counted the odds.

Luke was abnormal in all these things. His colleagues – he had few friends – considered him completely lacking in imagination. He just couldn't see himself as a ragged piece of bleeding meat or as a swollen or cindered corpse. He had in abundant measure the novice's supreme confidence that, never mind what happened to others, it simply couldn't happen to him. A pilot who served with Luke described him as 'the stupidest guy in the American Expeditionary Force, land, sea or air'. Yet America's top ace, Eddie Rickenbacker, said: 'Luke was the greatest fighter who ever went into the air.'

On the surface, Luke was an American 'tough guy' version of Ball and Guynemer, whose method – or lack of it – was to gain victories by attacking recklessly, returning, often as not, with aircraft shot to pieces. But whereas Guynemer was physically a weakling and had a 'timid' appearance, and Ball was just a nice boy who played the violin and had only one girl friend, Frank Luke had a horse, a pistol and cowboy gear, and when he was only seventeen he took on a touring professional prizefighter in the ring and knocked him out in the first round. He did not do well at school, but he was a first-rate all-round athlete in games, swimming and track, and a crack shot with rifle and pistol. He was not to kill nearly as many men as did Ball or Guynemer (let alone Richthofen), but that was because he didn't live long enough.

At twenty, Luke was 5 ft 10 inches high, weighed 170 pounds, and was hard and wiry. With a thin-lipped mouth and hard, grey eyes lacking in warmth, he would have made an excellent hero for a Hollywood western, smoking guns in hand, defying the 'baddies' or the Injuns. That is, in the 1930s. In the 1970s or 1980s, he might have starred as the mixed-up, rebellious youth without a cause, riding into the sunset on a souped-up motorbike.

Rebel he certainly was. All his life, he was utterly resentful of discipline. The trouble here was his father, who ruled his family of nine children with a firm hand. No doubt, with that many children underfoot, the father had to be hard to survive, but it distorted this son's character. What was to matter very much more, was that the father's idea of discipline was the sort of discipline imposed in the Prussian army – because that was exactly what the father had served in himself, as a conscript.

Frank had a German father; his country was at war with Germany. He would have to discover for himself who he was. And when he found out, it would not be surprising if he tried to prove himself twice as patriotic an American as the next man.

He took the unprivileged, uneducated man's road to aviation, joining the Signal Corps as a private soldier and putting in for flight training. In January 1918 he was commissioned as a second lieutenant. He received further flying training in France and was then posted to Orly aerodrome near Paris as a ferry pilot. In the Second World War, this was a job reserved often for women and over-age men. And that was exactly what Frank Luke thought of it. Furthermore, he said so. He had come over to fight, not ferry, he told anyone who would listen. Was there perhaps a doubt there? Was he given this noncombatant job because, being of German origin, he could not be trusted with a fighting aircraft over the lines? So, in his dark moments, Luke may have suspected.

Then, on 26 July 1918, all was resolved. Frank Luke was transferred without warning to the 27th Aero Squadron of the First Pursuit Group, flying Spads. The ferry organisation may simply have got rid of him because they were tired of his bragging. Anyway, the 27th knew before he came that the new man had a reputation as a malcontent and loudmouth. Luke immediately confirmed it. He asked them why weren't they all aces? They'd had enough time at the front, hadn't they? Luke himself had not been over the lines yet, or so much as seen an aeroplane with a black cross on it. To make matters worse, within days the whole group was embittered by heavy losses in air fighting. The 27th Aero Squadron alone lost six out of the seven machines they had put in the air, and that included the flight leader's. They had very dangerous enemies.

The CO Major Harold E. Hartney, included Luke in a patrol he led the next day, telling him to concentrate on keeping position in the formation and watching for his leader's signals. Luke was not interested. After half an hour he went off on his own. His excuse, when the patrol had landed just after him, was 'engine trouble' – a patently false story. He did the same thing on the next patrol. His story this time was that he had sighted a German two-seater and gone down after it. No one else had seen that two-seater and novices never could see other aeroplanes anyway – that was why, for their own sakes they had to be shepherded at first.

Luke was grounded for a week, as a chronic deserter, and he

retired sullenly into his shell. On his first patrol after the grounding, he again left the formation soon after take-off. This time Major Hartney told Luke exactly what he thought of him. The novice listened with a smile and then cut in: 'Anyway, I shot down a Hun.'

'The hell you did! Where?'

It turned out that Luke had very little idea of where he had been (which was normal for a novice) but could go into great detail on how he had attacked a complete German formation and knocked off the last machine in it. Hartley tried to get confirmation of the incident, but without success. The other pilots thought he was a liar, and shunned him. A journalist then covering the work of the Group later wrote in a book that he had had access to Luke's diary and it contained an admission that this first victory claim was false. He was certainly under very great pressure, and bitterly resentful at being made the outcast of the squadron. He was able to make one friend only – another pilot of German origin, Joe Wehner.

Wehner's plight was a good deal worse than Luke's. While Luke had a chip on his shoulder entirely of his own making, Joe Wehner was under active investigation by the secret service. Apart from his parentage, the only misdeed he had committed was to go to Germany for the YMCA and distribute mail and food parcels to British and American prisoners of war (these Americans had been fighting in the British or French forces, for America was not then at war with Germany). The moment America broke off diplomatic relations with Germany, Wehner returned home. He was understandably upset and sensitive about these suspicions and had also withdrawn inside himself. But he now struck up a quiet friendship with Frank Luke.

In his bitterness, Luke did not realise that both the CO and one of the flight leaders believed that he had potential and were giving him a lot of rope because of it. Wehner made him see this, and soon Luke got permission to do solo patrols (which the other pilots much resented). Throughout August and into September, Luke had his chances but produced no results. The only memorable thing he did was to fly a patrol with a new CO, Captain A. Grant, and break off from it to land and get drunk at the aerodrome of Guynemer's old squadron, the Storks. He drank so much he couldn't get back the same day, so was officially AWOL while on active service, not at all the same thing as whooping it up back in the States. Luke was grounded for the offence, and then repeated it. Hartney, now CO of the Group, stopped Grant from court-martialling this useless, drunken pilot.

On 11 September there happened to be an argument in the mess about observation balloons, in which the consensus of opinion was that they were much too dangerous to attack. No one would volunteer for such a task – he would have to receive a direct order before committing suicide in that fashion. They were much too well protected by AA guns and, closer in, by high-angle machineguns; and there was usually a small fighter strip nearby, with a couple of Fokkers on it. Luke was listening, fascinated. It was all new to him. He had never heard of the various aces, most of them dead, who on both sides of the lines had specialised for a time in balloon-busting. There was a technique to it, but he didn't know that. He simply said to Joe Wehner: 'I'm going to get one of those balloons. All those guys are scared of them. I'll show them.' He asked his friend if he would fly top cover for him next day, while he went down and burned a balloon, and Wehner agreed to do this extra patrol.

So off they went and circled the balloon, kiting in the sky on its long cable, with the observation basket below it. The observers had parachutes, and so did German pilots now. Luke's plan lacked subtlety. He just dived straight at the balloon, through a carpet of shell bursts first of all and then in the zone of crisscrossing machinegun fire which tore holes in the Spad's canvas. He blazed away until flame, tinged with smoke, appeared on the side of the balloon, then he dived for the American lines, landing not at his aerodrome but at an auxiliary field near to an American balloon unit, to get his victory confirmed. 'Did you see me get that balloon?' he demanded.

Meanwhile, Joe Wehner on his own, with no top cover against any Fokker patrol, had without fuss picked off the next German balloon in line. He did not land anywhere for confirmation, but simply entered the fact in his combat report. Luke, however, was ecstatic. 'I'm going to get more balloons,' he told the CO. 'I'm going to get a hell of a lot more balloons.' Grant promptly gave him one, which Corps wanted eliminated. Luke eliminated it, and another one within minutes, while a patrol led by Grant engaged eight Fokkers which tried to guard the balloons.

Luke knew so little that he believed he had discovered a hitherto unknown route to glory. He began to invent a technique (which others had thought of earlier, for it was obvious), of surprise attack. He suddenly thought: 'How about catching those bags at dusk when they haul them down? I'll bet half those guys on the guns will be having a beer or something.' It would mean coming back in the dark, and he knew little of night flying, but would chance that if Grant would set up flares. On 16 September, as he and Wehner were walking to their Spads, Rickenbacker and two other pilots were passing. Luke couldn't resist. Stopping Rickenbacker, and pointing to two far-off balloons hanging in the eastern sky, he told Rickenbacker: 'Keep your eyes on those two Kraut balloons. You'll see the first one go up in flames at exactly 7.15 and the other at 7.19.'

'Nutty as a squirrel,' remarked one of Rickenbacker's companions.

Then at 7.15 exactly a bright glow in the sky, growing steadily in size, appeared in the east. 7.19 came, and passed. But at 7.20 another tiny glowlight appeared and swelled.

'My Lord, he did it,' said Rickenbacker.

Luke's triumph was dramatically complete, because Colonel Billy Mitchell, head of the air service, had arrived at the aerodrome in time to watch his spectacular success and see his night landing by the light of flares. No one paid much attention when, much later, Joe Wehner returned. After covering Luke during his busting of the two balloons, Wehner had not followed his friend home. Caught up in the new game, he had tried his hand at the business again, and got two more. There had been no one, hanging above, to protect him from any Fokkers. Luke, without a thought, had gone for home as soon as he had got his own balloons.

Now he was bursting with confidence, eager to repeat the sunset trick, but was restrained by the quiet Wehner. Unlikely the Germans would be caught out twice, he said. 'Let's fool them. Let's start earlier and let's both go down at the same time. We haven't tried that yet.' So, using cloud cover to achieve surprise, the two Spads came at the first balloon of the first pair from both sides, shot it down and repeated the tactic with the second balloon. Both American aircraft were hit, but not fatally; the ground gunners were confused by the converging attacks. That had been easy, and there was a third balloon in the distance. This time there were Fokkers in the vicinity, so Wehner climbed towards them while Luke took the balloon.

This one proved stubborn. Luke had to dive three times before it burned. Then he saw yet another balloon to the north. Could he get that one, too? Make it four in a day? He turned for it. Above, trying to head off six Fokkers, Wehner fired a signal flare to warn Luke, who probably did not see it. The Fokkers were now between the two American planes and Wehner dived to attract their attention. The German formation split, three Fokkers to each Spad. Luke climbed towards his trio head-on and, with balloon-busting incendiary ammunition in his belts, shot two down in flames. Then he went home, without trying to find Wehner. On his way he saw a flight of Spads chasing a Halberstadt towards him, and he shot that down, too. It really was his day! But not Joe Wehner's. He was dying in a German hospital, with three bullets in him.

Luke wanted Wehner replaced as his top cover by a complete flight of Spads from the squadron. Captain Grant refused him this protection, unless the squadron had been actually ordered to shoot down a particular balloon. Some German balloons, he pointed out, were seeing just what the American army command wanted them to see, while others were useless. But Luke wanted personal victories and enlisted the help of a young admirer, Ivan Roberts, as top cover. That day, neither man even got as far as the balloons. They were jumped by five Fokkers. Luke claimed one Fokker, but Roberts went down behind the German lines. Luke's balloons were getting expensive. A balloon was cheap and easily replaced (not so the observers in the baskets, who usually jumped when an aeroplane came anywhere near them), but it would take a great many balloons to make up for the life of just one American pilot, and Luke had now spent two.

Probably, Luke suffered from remorse, for immediately after the loss of Roberts he went AWOL to the nearest town. Then, while he was away, the squadron was actually ordered to eliminate a particular balloon as a matter of urgency. But Luke was missing, absent without leave. One of the flight leaders had to do the job. When Luke came back next day and heard this, he took off in a rage and finding no balloons in the air, attacked one which was on the ground, partially inflated. It was military nonsense, and Grant grounded him for twenty-four hours, telling Luke that he was acting like a child.

Orders meant nothing to the rebel. No one would refuel his Spad, but he took off nevertheless and headed for an auxiliary field where he hoped no one would know of the grounding order. Grant found out, and had no choice. He telephoned the auxiliary aerodrome and told the flight leader there to put Luke under arrest the moment he landed. A court martial it had to be. But Luke was lucky – or unlucky, according to the way you look at it. The superior who had favoured him, Major Hartney, landed soon after him at the auxiliary field. Luke went over immediately, getting to the Major before the flight leader. 'Major, you're in time for a show. There are three balloons along the Meuse near Dun. I'll get all three, if you'll authorise the petrol.'

'Sure you can get them?' asked Hartney.

'Positive.'

Technically under arrest, awaiting court martial, Luke took off with full tanks, followed by Hartney. Even in those circumstances – or perhaps especially in those circumstances – Luke could not resist a grandstand play. He had scribbled out a note: 'WATCH THREE HUN BALLOONS ON MEUSE. LUKE.' Now he flew low over the American balloon headquarters and dropped it in a message cylinder. He was going to have an audience and confirmation.

The observers watched the distant German sky. One balloon glowed. Then a second. A long interval. A third glow shone out in the dusk and wriggled downwards.

But Luke did not come back.

The date was 29 September 1918. He had done it all in less than a month, and was dead. But just how Luke had died was not discovered until, after the war, a graves registration unit was checking at the town of Murvaux. Luke had been buried in an unmarked grave, and the Germans in this case had not notified the Americans of his fate. It turned out that there was a reason for that. Luke's Spad, so the French witnesses said, had come diving low over the town, pursued by German fighters, and flying erratically, after the balloons had been set alight. Some witnesses thought he had shot down two of the Fokkers. Anyway, he certainly had opened fire on German troops in the streets, killing some (figures varied from six to eleven). Given the special anti-balloon incendiary and explosive bullets he was using, their wounds must have been ghastly. By now, Luke was probably wounded himself, and the Spad damaged, for he landed in a field just outside the town, climbed out of the cockpit and staggered over to a stream for a drink of water. It was then that the German soldiers found him and called out to him to surrender.

It was not unknown for pilots to swear that they would not allow themselves to be taken prisoner, but when the moment came, of course they surrendered. Not so Luke, according to the French. He made a last stand by the stream, pulling out an automatic pistol and, apparently, shooting three Germans before one soldier raised a rifle and finished him. Luke fell forward on his face, just like any hero in a Hollywood movie.

In a movie, after the hero has died fighting to the last and the cameras have stopped rolling, he gets up again, dusts himself off, and lights a cigarette. But this was not Hollywood. Luke's body, stripped of its identity tags, was rolled into a shallow grave dug, on German orders, by two French peasants.

Even so, the story did not finish there, but ended as it often does in Hollywood. When the graves registration unit had completed their task of solving the mystery and the tale of Luke's last fight was known, it was Captain Grant, the officer who had so nearly court-martialled him, who recommended Frank Luke for America's highest award, the Congressional Medal of Honour. And it was Frank Luke Sr who took it on behalf of his dead son.

Chapter 5

THE HIGH SPEED FLYER

Sam Kinkead: 1928

On 12 March 1928 I was nine years old, recuperating from an illness at Ventnor on the south coast of the Isle of Wight. I wrote my autobiography shortly afterwards, completing it when I was twelve years old. In it there is a reference to that afternoon in 1928:

> One day a blue seaplane with silver wings passed overhead with a terrific roar. The evening papers told us that it was Flt. Lt. Kinkead trying for the world speed record held by the Italians, in a Supermarine S5. The machine had crashed into the Solent directly after we heard him and plunged straight to the bottom taking off the wingtips and poor Kinkead's head like a knife.

The titbit about decapitation was not in the papers. I had that from my father, who was a naval doctor, and he had it direct from the RAF doctors. High-speed flight was then novel, and producing medical problems of interest to both services – the tendency of pilots to 'black out' on a turn, for instance. My experience of medical people is that they love to explode their most stomach-turning stories at mealtimes, and father was no exception. He explained to my mother, my brother and me, that when the S5 was raised there was no sign whatever of the pilot. The cockpit was empty. So they presumed that somehow Kinkead had been forced out and was still somewhere in the Solent. They believed this until, while taking the S5 to pieces, the body of the pilot was found at the rear end, stuffed up the tail by the force of the water entering the cockpit, the result of hitting the sea at over 300 mph. This left a lasting and useful impression in my mind, that to strike water at speed would be like hitting concrete. By the time I was fifteen, I was flying aeroplanes myself and soon had the opportunity of inspecting one of our machines, a slow Cirrus II Moth, after it had dived in at Spithead. This was a wooden aircraft and it looked like a matchbox that has been crushed in the hand. That, too, was the effect of hitting water.

I seemed to remember that there was some mystery about the Kinkead crash; some part of it which was not entirely explained. When I decided for that reason to include the last flight of the S5 in this book of mysteries, I was also wary in case that part of the story which I had not written down, and which was entirely based on a memory of what my father had told me fifty-two years earlier, might be faulty. My confidence wavered when I read that the finding at the inquest had been that Flight-Lieutenant Kinkead had been killed when landing the S5. That is, at slow speed, not, as I seemed to remember, while he was making a high-speed run, the first part of which I had witnessed. Could I have been that much mistaken? If so, there could be no truth in what I thought my father had said about the body being forced up to the far end of the fuselage, for surely that graphically illustrated sheer speed of impact when the S5 hit?

The obituaries seemed incomplete. According to what was published at the time, Samuel Marcus Kinkead was thirty-one years old and a South African, educated at High School, Johannesburg, and at Marist College. During my inquiries, however, I discovered that Miss A.M. Kinkead, his niece, was living in Portsmouth, a few miles away from my home on Hayling Island. There was indeed a family resemblance. Miss Kinkead explained that her grandfather, Edward Kinkead, was a Scot from Stirling who had gone to Jamaica and started a business. He had married there, one of his sons being S.M. Kinkead. Then his wife had died, he had left Jamaica for Southampton in England, and married again, having two more children, one of them being Miss Kinkead's father, who became chaplain to St Mary's Hospital, Portsmouth, and was the half-brother of Flight-Lieut. Kinkead. Because the families split up, Miss Kinkead could not remember Sam Kinkead well, but she recalled wearing a seaplane brooch during the Schneider Trophy contests, two of which were staged at Portsmouth, in 1929 and 1931. That brought it back to me, because I still had a similar rosette from the 1929 contest, which I had described in my 'autobiography'. It was a thrilling affair, with the Italians participating, whereas 1931 had been just an all-British walkover. Even so, one of the pilots had been killed, because flying these high-speed racers allowed no margin for error or ill fortune.

There was nothing like them nowadays, so if I was to understand how they were built and flown, I would have to find some of the very few people who had been at the heart of the story. I was lucky. In 1929 I had made a brief diary entry; 'While practising for the Schneider Trophy, D'Arcy Grieg came very low over the South Parade Pier, which is just in front of our house, and the cat dived behind the sideboard at the sound!' I discovered that D'Arcy Grieg, now an Air Commodore, was living in retirement in Sussex, a couple of hours' drive away, and that, on the death of Kinkead, he had been the officer selected to form a new High Speed Flight and take the world record from the Italians in another Supermarine S5 seaplane.

It was fortunate also that Southampton enthusiasts had founded a museum to commemorate R.J. Mitchell, the man who designed both the S5 and its derivative, the Spitfire fighter of the Second World War. In the museum, in addition to a Spitfire, an S6A was on display. Larger and heavier than the S5, it was still sufficiently like it to give a fairly horrifying idea of what it had been like to sit inside one of these dangerous rockets. I knew well what they looked like in action, for I had watched the 1929 contest from underneath the course just west of South Parade Pier. There was no warning at all. The machine – blue and silver if it was British, blood-red if it was Italian – suddenly appeared over the roof of the pier in absolute silence, shot overhead trailing exhaust smoke at a height little above that of the seafront hotels, and only as it reached Southsea Castle a quarter of a mile or so away did we hear 'a shattering whining roar', the sound of the engine apparently following 300 yards behind the aircraft.

The highlight had been when one of the Italians, I think it may have been Del Monte, blinded and in agony from hot oil splashing into the cockpit, had almost brushed the pier roof with his floats, then swerved partly out of control towards the Savoy Hotel behind me, before correcting at the last second. At speeds of over 300 mph and at heights of under 300 feet, a fractional error could cause a spectacular accident. I thought the pilots were heroes and still possess souvenir postcards of the men and their machines. I never thought I would meet one.

Kinkead had a quite extraordinary number of decorations – five British, three Russian. He joined the RNAS in 1915 and first saw action at the Dardanelles. Then he was with No. 1 Squadron in the Dunkirk area, flying first Sopwith Triplanes and later Sopwith Camels. This squadron specialised in dangerous ground strafing attacks, at which Kinkead became an expert; on one occasion, he shot up German troops in a wood while being attacked by six German aircraft. By the end of the war he had scored thirty victories, ranking twenty-third on the British list of air fighters. He now had a double DSC and a double DFC. To these he added the DSO and the Tsarist medals when he flew with No. 47 Squadron supporting the 'White' forces in South Russia. The DSO was for leading a ground attack by a Camel formation on the 'Red' cavalry division of Dumenko at a critical moment in the defence of Tsaritsyn.

Engine failure over the Bolshevik lines often meant an unpleasant death. This was part of the risk also when Kinkead was serving in the Middle East with No. 30 Squadron a few years later. Ira Jones, who had served on the North Russian front, was with the same squadron at this time, operating against Kurdish rebels in Iraq. Twice Kinkead landed under rebel fire in order to help pick up the crews of British machines which had been forced down, and on ground which was rough and likely to wreck his own DH9A. In both cases he managed a successful take-off with a badly overloaded bomber. 'When he went deep into the Solent in his Supermarine S5.' wrote Ira Jones, 'the Royal Air Force lost, without doubt, its finest junior officer.'

The final stage of his career began when Flight-Lieutenant Kinkead was posted to Felixstowe for duty with the new High Speed Flight. The contests had started in 1913 with the presentation of a beautiful trophy by Jacques Schneider to improve marine aircraft. Three victories in succession would give the nation concerned permanent possession of the trophy. 1922, when the contest was won by a Sea Lion flying boat built by Mr Pemberton Billing's Supermarine firm at Southampton, marked the last success by a private entry. From now on, only nations could afford the outlay on racing teams and the development of new, fast machines for them to fly. The three which cared to do so were America, Italy and Great Britain, and the machines so produced really were excitingly new, at the frontiers of existing technology.

In 1923 a neat-looking biplane, the US Navy Curtiss, won at

Cowes in the Isle of Wight at an average speed of 177 mph. In 1924, America generously declared the contest void, as there were no other entries, and in 1925 won again with a Curtiss biplane at a most convincing 232 mph. The British had two modern-looking entries, the Gloster III biplane and the Supermarine S4, a stylish mid-wing monoplane. The latter crashed, it was said on alighting, but the cause may have been wing-flutter. At this stage of development, the biplane was markedly stronger than the monoplane, but it was a monoplane which won at Hampton Roads in 1926, a Macchi M39 flown by Mario de Bernardi for Italy at a speed of 248 mph (although one of his comrades was killed in another Macchi).

The next contest was to be held at Venice in 1927 and it was to compete in this that the RAF High Speed Flight was formed. The Air Ministry ordered seven racers, and six were sent to Venice – one Short Crusader monoplane with 800 hp Bristol Mercury radial, two Supermarine S5 monoplanes with 900 hp Napier Lion engines, and three Gloster IV biplanes with Napier Lions. On its first test flight the Crusader crashed spectacularly on take-off, apparently trying to do a half-roll fifteen feet up, and dived into the sea at about 150 mph (240 kph); the wooden fuselage broke at the cockpit and the pilot, Flying Officer H.M. Schofield, was thrown clear, severely bruised. During rigging at Venice, the aileron controls had been crossed; the machine therefore reacted in precisely the opposite way to that which the pilot expected.

Three machines from each country were chosen for the final contest, the Italians being Macchi M52s, but only two out of the six finished – the S5 flown by Flight-Lieutenant S.N. Webster at 281 mph (453 kph) and the S5 flown by Flight-Lieutenant O.E. Worsley at 273 mph. Engine trouble eliminated all the Macchis and also the Gloster, but Kinkead did extremely well on the slower machine, actually overtaking Worsley's S5 before he was forced down by a cracked airscrew boss. The engines were being boosted by all sorts of means to give great power for a short time and because they had to fly a triangular course of fifty kilometres, the sharp turns produced average speeds well below the actual straight-flight capabilities of the aircraft. When, shortly after the Venice contest, in November 1927, Colonel Mario de Bernadi flew a Macchi all-out, he obtained a new world's speed record of 297 mph (479 kph). The magic 500 kph was now very near, let alone the round 300 mph for the Anglo-Saxon world!

Early in 1928 the High Speed Flight was reformed around the reserve Supermarine S5, N221, with Kinkead as pilot and Flying Officer T.H. Moon as engineer officer. Southampton water was chosen for the attempt because it is virtually enclosed by land – the coasts of Hampshire and the Isle of Wight – thus giving an area of protected water for take-off and alighting which the tricky seaplanes required. On a flat calm they would not take off at all, or only with great difficulty, and in much more than a breeze they were likely to crash, while a swell could be extremely dangerous.

The regulations of the International Aeronautical Federation stipulated a three-kilometre course, to be covered twice in each direction at a constant height of not more than 150 metres. Five hundred metres before entering the measured three kilometres course, the machine had to be at the height of 150 metres. This was to prevent pilots building up excessive speed in a dive, but, within the regulations, what they did was to gain height and then dive, aiming to level out at 150 metres approximately 500 metres before the measured run began. In the context of Southampton Water, that meant a dive from high over Romsey followed by a dive from high over the Isle of Wight, or vice versa, the measured three kilometres being between Calshot and Fawley. It was the start of the dive from over the Isle of Wight which I saw on 12 March 1928.

The team had been hanging around for some ten days, in perfectly hopeless weather which had allowed only one trial flight – a successful one, however. To familiarise themselves with the sea area, the pilots were taken for rides as far as the Nab Tower in the 38-knot Coastal Motor Boat (CMB) which was the fastest of the RAF rescue craft. As it had no astern gear, its ability to pick up a man in the water was problematical, but it could certainly go. Mr T.L. 'Monty' Banks, then a corporal engineer in the RAF, was in the crew of this boat both in 1927 and 1928. He recalls taking Kinkead out on a high-speed run in it and the pilot exclaiming: 'My word, that was a thrill!' as the boat slowed. Banks was surprised and remarked: 'It can't be much of a thrill after the speed you've been used to travelling around.' Kinkead insisted that it was. 'On this thing, you can see your speed. Up in the air you cannot.'

There was no 'side' about Kinkead. He was modest and quietly spoken; from his manner, no one would have guessed at his war record or that his present task was at all out of the ordinary. On the morning of 11 March, which was a Sunday, after the attempt had

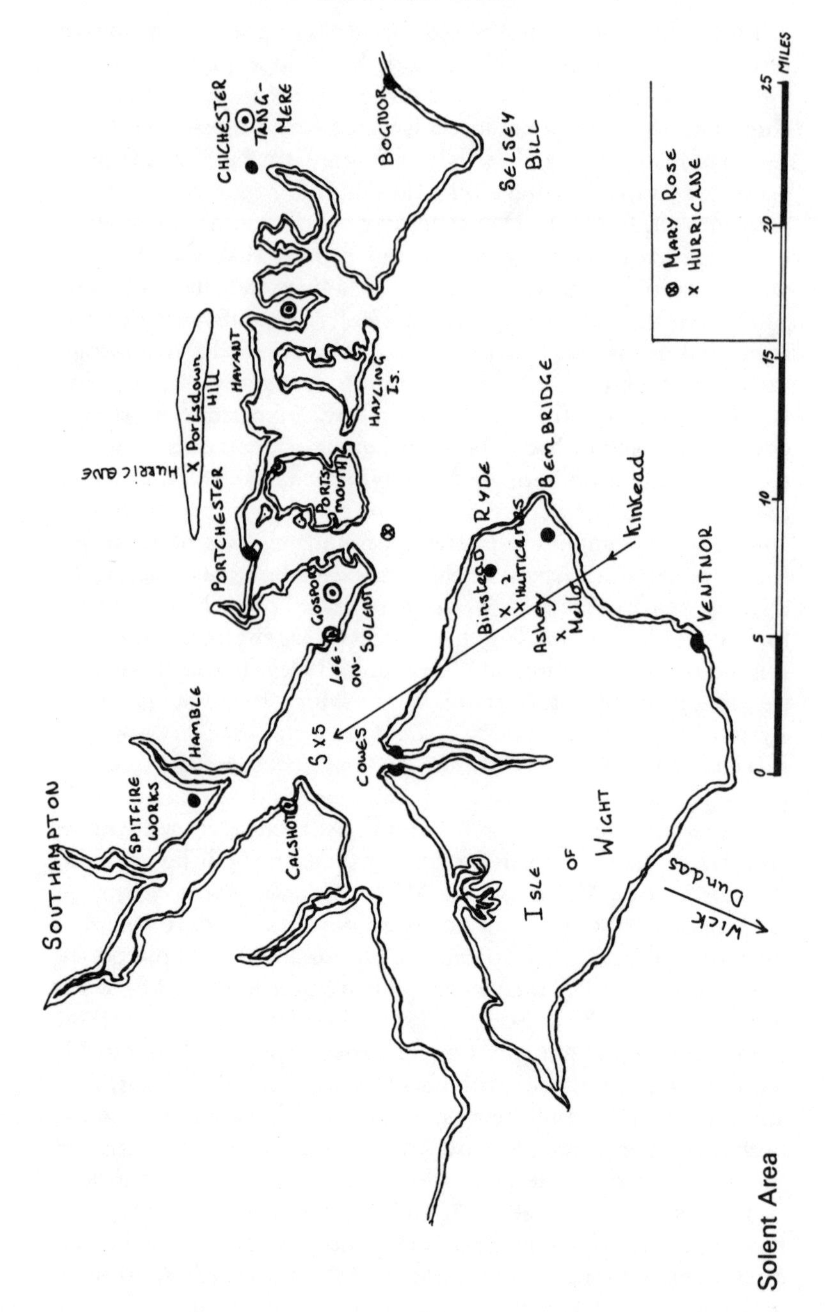
SOUTHAMPTON
SPITFIRE WORKS
HAMBLE
CALSHOT
COWES
5 x 5
LEE-ON-SOLENT
GOSPORT
PORTCHESTER
PORTSMOUTH
X Portsdown Hill
HURRICANE
HAVANT
HAYLING IS.
CHICHESTER
TANG-MERE
BOGNOR
SELSEY BILL
RYDE
BEMBRIDGE
Binstead
X 2
X HURRICANES
Ashey
X
Mellor
Kinkead
VENTNOR
ISLE OF WIGHT
Wick
Dundas
⊗ MARY ROSE
X HURRICANE
0
5
10
15
20
25
MILES

Solent Area

been cancelled once again, Kinkead came aboard the CMB through the forward hatch and introduced himself (not to the crew, who knew him, but to the new skipper of the boat) by saying: 'The team call me Sam,' adding: 'Well, given favourable weather, this job will soon be over. I was hoping to do it today.' As *Flight* had just pointed out (8 March): 'At the speed at which Kinkead will be flying a lap will take about 20 seconds!' Put another way, the S5 would be travelling, a hundred feet up, at a speed of 450 feet per second.

Next day, 12 March, the weather seemed undecided at first. But the CMB 'Monty' Banks was serving in had to leave its base in Haslar Creek, Portsmouth Harbour, at four in the morning in order to be in position on time. The watchkeepers on Calshot tower used to say that they could hear the CMB, with its powerful, unsilenced engines, actually leaving Portsmouth Harbour, and knew they could clear the aircraft for launching. There was not supposed to be any flying until all the rescue craft were ready and in position. In addition to this very fast boat, Calshot had to provide a slower rescue launch which, however, did have an astern gear. This was known as the 'Duty Stand-by Flying Boat'. William Charles Etherington, also an RAF corporal and a friend of Banks', was with this boat.

There was a third witness, still living, who was then on the water also. He was Mr D. Pragnall, a crew member of a vessel owned by the James Dredging Company of Southampton, which was deepening the channel off Hamble jetty for the oil tankers. At the critical time he was to be in a rowing boat and about to refill the lamps of the buoys laid out to mark the dredging area at night. A distinguished audience, including the S5's designer, R.J. Mitchell, and various foreign Air Attachés, had come to witness the speed attempt, but in the event it was the witnesses on the water who were to be closest to the scene. Oddly, they were not called on to give evidence afterwards.

Events seemed to conspire against the attempt. Early in the morning when conditions seemed favourable, an oil leak prevented the S5 from getting off. When that was cured, the weather had broken. Soon it was snowing. About mid-afternoon, recalled Banks, flying was cancelled for that day and the CMB returned to Haslar in the most blinding snowstorm he had experienced, running on one motor throttled to dead slow and making 'plenty of music on the klaxon'. They hurried into the mess, hoping for early tea and a run ashore. It was not to be. A message came, telling them to go back to Calshot. 'We couldn't understand it,' recalled Banks:

> We thought the weather was totally unsuitable. But we had to go. Up to that time, the machines were not allowed to take off until we were in position. So the skipper went hell bent for Calshot, not knowing that the S5 had already taken off. The deck staff didn't know he was in the air until they saw him. The skipper slowed when he saw the machine and, as I was inquisitive, I stood on the starboard engine bearer and had a look through my little peephole.

Corporal Etherington had watched the take-off shortly after five o'clock that afternoon. The snowstorm had passed and the sea was now very calm, but conditions were still doubtful. 'They were dubious about taking off because a bit of a sea mist was coming up,' recalled Etherington. This could mean that the pilot would have no visible horizon by which to judge the attitude of his aircraft, and he might also find it hard to judge his height above water. But Kinkead, wearing a white sweater, got into the cockpit of the racer. The machine had been designed to produce the smallest possible frontal area – no higher than the height of the Napier Lion engine. Consequently, there was no room for a seat – the pilot sat on a cushion placed on the floor of the fuselage with his legs straight out in front of him. There was no belt or harness with which he could strap himself in. The cockpit was so cramped that he had to wriggle in sideways, and then, when he had straightened up, his shoulders were jammed underneath pads fitted inside the fuselage which held him in place. Sitting in the S5 was like wearing a tight-fitting suit of hot metal clothing.

The cockpit was open, but the pilot's head could not be allowed to stick up above the line of the fuselage and break the airflow, so he was seated so low that all he could see directly ahead was the engine. He could look forward and up, of course, and forward to one side or the other of the Napier Lion, but there was no view at all straight in front. Sideways, the view was quite good, except on take-off, when for a short time nothing at all could be seen, ahead, above or to the side. As the seaplane surged forward under the tremendous power of the Napier Lion, creating a bow wave under the floats, the propeller sucked up the water and gave the pilot an almost solid showerbath. At the same time, as soon as the throttle was opened, the flailing propeller created tremendous torque, forcing the seaplane down on one side and practically submerging one float until it lifted onto the step. After that, according to D'Arcy Grieg, the S5 took itself off the water and, to fly, became just like a fighter aircraft. There were snags. If the seaplane developed a porpoising motion on take-off, the throttle had to be closed instantly, otherwise it would bounce into the air, stall and dive in (Lt. Brinton RN was killed in 1931 in an S6 in this manner). The landing approach took several miles, because these machines were not fitted with flaps or variable-pitch propellers, later inventions to slow down an aircraft for easy landing. The S5 alighted at about 90 mph and it had to be on the tail-ends of the floats, not on the steps, so a very flat glide at 120–130 mph was required, with the pilot holding off, holding off, until finally the seaplane settled down with its nose in the air. There was no way a witness could confuse an attempt on the world speed record with a landing, provided he had a good view.

Not everyone did, of course. Some, like Mr L. Bartholomew heard the plane in the air, then the sound of the engine ending abruptly. He never actually saw the S5. Others saw Kinkead take off with difficulty, because it was very calm, circle round, make a landing (as required by the regulations) and, taking off again, disappear towards Cowes in the Isle of Wight. Mr R. Sevier was building a beach hut that afternoon, off Hurst Castle, and had as good a view of the crash as the misty weather permitted; he had the strange impression that the engine continued running even after the seaplane had struck the water. Mr A.G. Munnings, who was then a young apprentice at Fairey Aviation's works on Southampton Water, helping to build Fairey IIID seaplanes, came outside to have his tea break just in time for a grandstand view. It never occurred to him that Kinkead was trying to land. Quite obviously it was a high-speed run. But there was nothing dramatic about its end. 'I can explain it like a man about to swim underwater,' he said. 'There did not appear to be a great splash. I expected him to reappear.'

For many people the most horrible aspect about Kinkead's end was the way in which, because the sound travelled behind the machine and it was approaching Calshot from Cowes, the high screaming of the engine continued to sound above the quiet water for a little while after the S5 had disappeared.

'Monty' Banks, of course, heard nothing at all. No one did on the CMB, because of the overwhelming howl of its own engines, but, peering through the perspex peephole from the engine room, he saw the seaplane dive, and then he was sure he saw the tail flutter just before the machine hit the water. The skipper opened up both engines to full and the CMB roared towards the scene of the crash. They arrived within minutes, but two other boats were before them.

Mr Pragnall had been sculling the dredger's small boat away from the ship towards the first lantern buoy when he heard the S5 coming and looked round. 'He was so close to the water, I said to meself: "He's not going to make it." And this time he dived right in – perhaps half a mile from where we were.' Jack Unwin, skipper of the dredger, called Pragnall back to the ship and got into the boat with him, saying: 'Pull out quick.' 'We were actually the first at the wreck,' Pragnall remembered, 'but we couldn't see any body. Just bits of wings about, on top of the water. We thought we'd help the pilot, but RAF motorboats came out to the wreck and told us this was their business.'

The first RAF boat to arrive and buoy the wreckage was the duty launch in which Corporal Etherington was serving. 'It was very quiet, no wind. The mist was not too bad when he took off. Patchy. But a sea mist comes down quick. It was hanging above the waves as he was diving to the measured course – and he just went straight through. I saw the actual crash. He didn't flatten out at all. I couldn't see the splash for sea mist. It was all over in a flash. We heard him coming, he disappeared, that was it. No sign of wreckage, but we knew where he'd crashed and there was an oil patch. So we went over and dropped a buoy on it.'

The CMB came roaring up next, behind a great, curling bow-wave, with the senior deckhand, Corporal Gallimore, about to dive overboard to the rescue. He had taken off his sweater and his boots. But the skipper put a hand on his shoulder and said: 'No good, corporal, he's gone right to the bottom.' There was oil in the water, recalled 'Monty' Banks, but hardly any wreckage. The crash had occurred in moderately deep water near Calshot lightship and it took several days before a salvage vessel had located the wreckage, which was in two parts, and had lifted it to the surface.

Banks remembered seeing the divers at work, and the great liner *Leviathan* passing up to Southampton and dipping her ensign in salute.

The remains of the S5 were taken to Calshot and the controls laid out on the slipway to check for a technical fault such as the crossed controls which had nearly killed Schofield at Venice the previous year. But the Inspector could find nothing technically wrong with the machine.

Two people, Sergeant Temple and Corporal Etherington, had the job of getting the body out. I had rather wondered how accurately I had recalled what my father had told me more than fifty years ago, but it turned out to be quite true. 'The head was off and the body was up the tail. Sergeant Temple and I had to cut the tail open to get the body out. It had been cut as if by fine wire. Half his head had gone – it was a vertical cut, more or less.' Mr Etherington gave a grimace of distaste. 'He was wearing a white jersey, but it had been turned to jam and then, remember, the body had been under water for two days, after he'd hit at 300 mph. I was a bearer at the funeral, and I didn't like being a bearer, knowing what was in there. I thought about it all the time, remembering how he was when we got him out.'

I received detailed confirmation in a curious way when trying to trace the people who usually managed to be left out of the reckoning when praise was being distributed for the success of the Supermarine seaplanes – the men who actually built them. Mr E.W. Gardener wrote to me. He was the very man who had dealt with the wreckage when it had been returned to the Woolston works of Supermarine. 'The wing on impact with the surface of the water snapped off and caused the bulkhead to be pushed along the fuselage, breaking the 1″ × ¼″ steel flying wire which was bolted to the top of the bulkhead. I should have a piece about 18 inches long among the junk in my shed, unless some blighter's pinched it.' Mr Gardener, who had also worked on the early Spitfire, died before I could visit him.

One thing was quite certain. Sam Kinkead had died instantly. But why? Two inquiries were held simultaneously at Calshot: the RAF official inquiry began on 14 March; the Coroner's Inquest opened the following day. Curiously, the first expert witness called before the Coroner, the Adjutant of the Calshot base, had not witnessed the accident; he had merely heard the S5's engine stop.

He could not estimate how fast the machine was going at the time. He could not say whether Ethyl Spirit was being used. He did not know if Kinkead had been affected by fumes on previous flights from Calshot. He could not say if Kinkead had suffered from fumes during the Schneider Trophy contest at Venice. He could say only that Flight-Lieutenant Kinkead had no brother (and there he was mistaken).

The Coroner pressed the points about fumes with almost every witness, many of whom were reluctant to answer, probably because the fuel used was a technical secret, employed to boost power for a short period. Some admissions he did obtain, however. Flight-Lieutenant Cundlewood said he had spoken to Kinkead after his practice flight on 11 March; Kinkead told him that he had been sick that morning, but now felt perfectly well. R.E.G. Smith, Napier's service engineer, said that they were using a special fuel not in use elsewhere. He had known people to be affected by the fumes from it, but 'the deceased was the only man I have known who has been made sick by the fumes.' It was possible that he had been overcome by the fumes. And yes, he had heard Kinkead say, at Venice the previous year, that 'he felt the effect of the exhaust fumes in his stomach to a minor extent for a period not exceeding twenty-four hours after any particular flight, but not necessarily every flight.'

The Coroner, Mr P.B. Ingoldby, adjourned the inquest in order for samples of blood and the right lung to be sent to the RAF Pathological Laboratory at Halton. When the inquest was resumed on 26 March, a very long report was read out by an RAF doctor: exhaustive tests had disclosed no trace of poisoning by carbon monoxide or by lead or a lead compound; there was no evidence that exhaust, or petrol, fumes had affected the pilot.

The inquest concluded on 20 April with the evidence of Major J.P.C. Cooper, OBE, MC, Inspector of Accidents for the Air Ministry, who stated that there was no evidence of technical failure of either the air frame or the engine. When asked if evidence of tail flutter had been given at the RAF inquiry, Major Cooper answered 'Yes.' In his view the witness had been mistaken: what he had seen might have been the reflection of sunlight from the rudder as the machine turned. 'Monty' Banks did not like the way in which the suggestion of flutter was brushed aside; the witnesses interviewed had been on the seawall at Calshot, much further away than the crew of the CMB. Yet their skipper had not been called to give evidence.

Asked if he had formed an opinion as to the cause of the accident, Major Cooper replied that he had. 'The aircraft stalled at a height of something like fifty feet to eighty feet. I formed the further opinion that the pilot misjudged his height above the water when attempting to land, and this led to the machine stalling. I think the pilot abandoned his attempt to fly the speed course because of the weather conditions at the time.' Major Cooper explained that he meant the mist, the lack of wind, the calm sea, the lack of horizon. The Coroner accepted this view and returned a verdict of death by misadventure.

A major mystery is Major Cooper's belief that the S5 stalled onto the water when landing, as everyone else described a descending high-speed run to the measured three-kilometre mark. The witnesses in the boats, being to one side, had the best view, but even those who were watching Kinkead coming at them almost head-on were in no doubt that he was travelling extremely fast. R.E.G. Smith, the Napier engineer, told the inquiry earlier:

> The machine had made one flight and a landing on the water immediately prior to the flight which ended fatally. The machine took off from the water, made a wide circle to the left, and disappeared in the direction of Cowes. The machine was out of my view and hearing for about three minutes. I was standing on the eastern slipway when the noise of the engine again became audible. The machine soon appeared in sight, travelling towards me from Cowes at about 250 feet above the water and going very fast. It was impossible to tell the pace. The machine appeared to be intentionally descending, and after a few seconds, nose-dived very suddenly into the sea from about fifty feet up. This happened about a half to three quarters of a mile away.

The landing speed of the S5 was about 120–130 mph, the stalling speed 90 mph. At the stall, of course, the machine would simply cease to fly and then drop into the water. Both *Flight* and *The Aeroplane* referred to a speed of about 300 mph and considered all possible causes in that context.

Immediately after the accident, C.G. Grey listed a number of such causes. The first was tail flutter, caused possibly by the water thrown up by the propeller on take-off weakening the tail unit so that 'it collapsed when the machine was forced to full speed on approaching the timed course.' Possible, but improbable, he thought. The S5s had been well tried at Venice and besides 'Mr. Mitchell is the kind of engineer who takes no chances.' Then he considered the possibility of a waterpipe or oilpipe freezing, or even the lubricant in the hinge of a control surface – for it was late afternoon in March and there had just been a heavy snowstorm. Even a partial stiffening of the controls at that speed and at that height could bring disaster, he thought. 'Apart from that, there are the possibilities of something happening to the pilot himself – ice on the screen in the cowling, or ice on his goggles might blind him, or even the glare of the sun, low on the horizon.'

A writer in *Flight* mentioned one theory, based on a witnessed speed of 'well over 300 mph' at about 100 feet, which meant that the machine would always be only one second away from the water. 'Thus even a momentary disturbance of any sort might well have caused the disaster. It has been suggested quite seriously, that if the pilot were to sneeze, this might in itself cause him to depress the elevators sufficiently to make the machine swoop down into the sea.' Variants of this theory were still being suggested fifty years later, mostly based on the experiences of Schneider Trophy pilots afterwards. One pilot, for instance, nearly crashed because one of his shoes came off and jammed the rudder bar. Another got cramp in a leg, rested it against the side of the cockpit – which was really a radiator – and was agonisingly burned. I myself saw the sudden swerve made by the Macchi over South Parade Pier, in 1929. I felt that no one could ever prove 'pilot error' in the case of Kinkead, which must always hold a slight tinge of mystery.

Wing Commander A.H. Orlebar, who led the High Speed Flight at one time, was to write:

> I do not know just what happened, but they had been forced to hang about for a long time for weather, and when the attempt was made the weather was not good. From a photograph of the machine taxiing out one can see it was certainly very misty and the water was very glassy. People say Kinkead was not too fit at the time, and the combination of difficulties may have led him to make a mistake. However, there are many different opinions, and many spectators believe that something went wrong with the machine.

D'Arcy Grieg was closer to the accident than Orlebar, for he was the pilot selected to fill the dead man's shoes and form a new High Speed Flight, based on the S5s. They were treated like eggs, sparking plugs removed after every flight, on principle, and engine stripped and inspected every five flying hours. Grieg was a fighter pilot and, as he said, seaplane experience did not really count – the racers handled like fighter aircraft with floats on them, and compared well to the Spitfires and Hurricanes which came later. He made thirty-four flights in racing seaplanes – Supermarines and Glosters – but his total flying time on them amounted to 11 hours 23 minutes. No one ever built up much time on these machines, but Grieg had more than most. The finest of the lot, in his opinion, was not a Supermarine at all, but the beautiful Gloster 6 which stalled at 110 mph. Unfortunately, engine trouble which could not be cured in time for a contest put this machine out of the running, and it was won by an S6, a heavier and more powerful version of the S5.

Grieg was in no doubt at all as to the cause of Kinkead's accident. There was no question of mechanical failure, but instead a combination of various circumstances which proved deadly:

1 Malaria. Kinkead was recovering from an attack, and so was a bit below par.
2 It was March and late in the afternoon.
3 There was a glass-calm sea which invites disaster, because unless there are objects floating on the water it is impossible to judge height accurately.
4 There was a mist, so that he had no horizon when coming in.

The calm sea and the mist combined would be enough to explain the accident, thought Grieg, without adding the other factors. He had himself, during a long flight in the S5, felt as though he was being suffocated by the fumes from the engine in addition to the heat generated by the oil coolers which turned the cockpit into 'something approaching an extremely hot Turkish bath'. And he was deafened by the noise of the 900 hp engine just in front of him.

In these conditions, and with the pilot suffering from a recurrence of malaria, it might be all too easy to make a mistake. But why then did Kinkead take off in the first place? The delays and the pressure on him to beat Bernardi's record would be one factor, but I think the major clue lies in Corporal Etherington's account. During our talk I got a very good picture of the weather conditions he was describing – the mist arising suddenly and clinging to the water in patches, so that there is visibility here but no visibility a hundred yards away. I had been going out in the Solent myself in small boats, on diving expeditions, for more than twenty years, and I recognised the sort of conditions he was talking about. In a

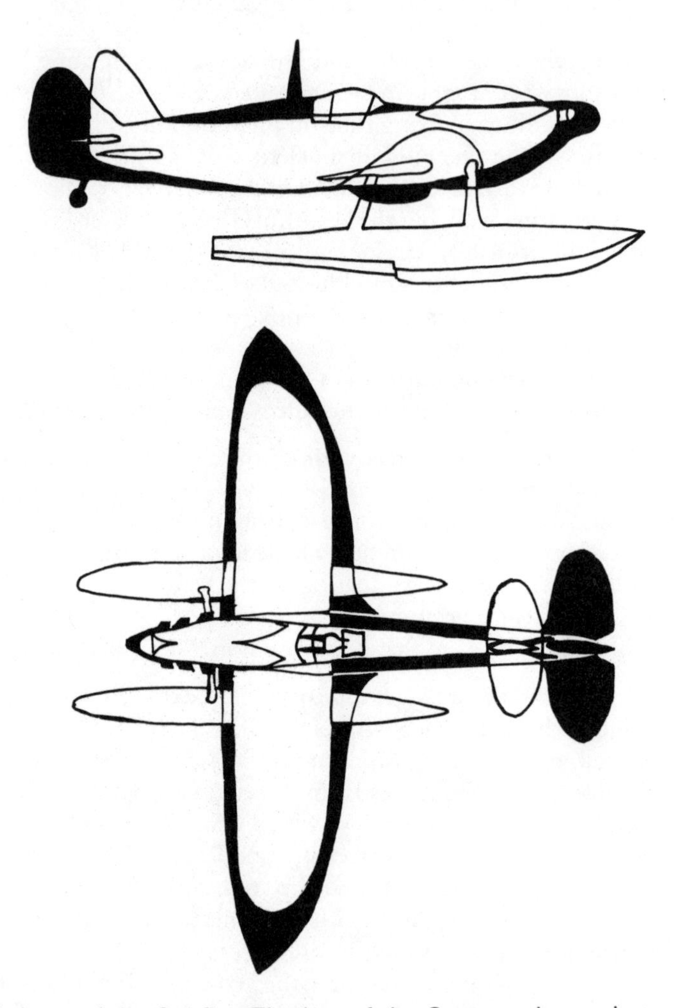

Lineage of the Spitfire: The last of the Supermarine racing sea planes, the S6B (white) superimposed on the Spitfire I (black).

small boat, they frighten me. A racing seaplane, once aloft after the mist had moved in, would be trapped. There would be no safe way back.

My talk with Mr Etherington was on 8 March 1980, fifty-two years later nearly to the day. On his directions, I easily found Fawley Church and the grave beside it. But now, Samuel Marcus Kinkead had companions. There was a neat row of RAF war graves a few feet to the north. Nine men had a single date inscribed – 15 August 1940. That was the greatest day of the Battle of Britain, which saw the largest forces engaged, the heaviest fighting. A tenth man had 16 August inscribed – that was the day they dive-bombed all the aerodromes, with balloons down in flames everywhere. And an eleventh man had not survived 30 September 1940, a day when the Luftwaffe had gone for West Country targets mainly. This was what it had all been in aid of – the Schneider contests, the attempts on the world record – this was what was coming a dozen years after Kinkead's death.

Too much can be made of the connection between the Supermarine racing seaplanes and the Spitfire, but a number of lessons were learned and used in the design of the fighter and the Rolls Royce engines which powered the S6 series were developed into the Merlin fitted to both the Hurricane and the Spitfire. Great Britain did win three Schneider contests in a row, and keep the trophy, but that hardly mattered compared to the development of a fighter which could match the Messerschmitt 109. A great deal more than the vindication of the Montrose Ghost is owed to Mr Pemberton Billing and his Supermarine Aviation company. It is interesting to ponder what might have happened in 1940, if the aeronautical empire builders of 1916 had had their way and eliminated him and his company. They would have eliminated the Spitfire as well.

Perhaps the strangest thing of all is the headstone just behind Kinkead's at Fawley, marking the resting place in English soil of

Flying Officer R.H. Immelman
(of South Africa)
Pilot, RAF, age 24
Killed in Battle of Britain

Chapter 6

VANISHED

The Lost Heroes and Heroines: 1910–1941

A man of Newquay in Cornwall was walking the tideline when he saw a large piece of wreckage – the complete rudder of an aeroplane covered in silver-grey fabric. The Channel tides might easily have brought that in from the Atlantic, so he told the coastguards. Later that day they found part of an aeroplane's wing of the same silver-grey colour. Some twenty aviators of different nationalities had disappeared in the last six months, all of them without trace, so the news of this discovery was wired to London, Paris and New York. Which one of those mysteries was on the way to solution?

As it turned out, none of them. When Air Ministry experts examined the wreckage, they identified it as belonging to an old aircraft which had been used for target practice off Plymouth, up-Channel from Newquay. Yet another false alarm. For the time was September 1927; six months and many rumours had passed by since Nungesser and Coli had taken off in the white-painted Levasseur seaplane *Oiseau Blanc* (*White Bird*), never to return.

Captain Charles Nungesser was the complete opposite of Georges Guynemer. He was a tough, Commando-type character, an ex-Hussar who in the mobile warfare of early 1914 had infiltrated German territory, ambushed a staff car, shot the occupants, and driven the captured German vehicle back to his own army's positions under a hail of French fire. He was an outstanding swimmer, boxer, motorcyclist and racing car driver; the first time he got into an aeroplane, he flew it solo, and – unlike so many other future aces – he did not break it. He enjoyed scaring people witless by low-flying and, flouting superstition, he had every death-sign he could think of painted on his fighting scout – a coffin, lighted candles, skull-and-crossbones, a black heart. In crashes and combats Nungesser amassed a crippling toll of injuries which kept him in hospital for months at a time; but he always went back to the front before he was healed, even if it meant struggling out to his Nieuport on crutches, or with the aid of mechanics. By the end of the war he had scored forty-five victories.

For such a man, peace was boring. He planned to make a most impressive Atlantic flight – Paris to New York – which was longer and more difficult than flights from Newfoundland to Ireland aided by the prevailing wind. A number of people had made that flight; none so far had succeeded in the westbound crossing. The ace and his colleague, Captain Coli, took off in the *White Bird* on 8 May and lumbered westwards at about 300 feet, so heavy was the load of petrol on board. That was at 5.18 am. At about 10 that morning the plane was sighted and identified off Ireland, heading out over the North Atlantic. And that was the last definite check.

Thirty hours later, the rumours began. The French plane was reported first over Nova Scotia and then over Newfoundland, on course for New York; and France went wild with triumph. It was premature. Nungesser and Coli did not land at New York. They did not land anywhere. Other planes had been mistaken for the *White Bird*. In one case, two responsible witnesses in Newfoundland had heard an aeroplane pass over Harbour Grace at the right time and on the right course to be the French plane; but it was in mist and they had not seen it. There was also a conflicting report of aeroplane wreckage sighted off Newfoundland. At the same time, from the other side of the Atlantic, on the west coast of Southern Ireland, there were many witnesses who claimed to have either heard or seen an aeroplane dive into the sea. This was why the finding of silver-grey wreckage in the western entrance to the English Channel had caused so much excitement; had it been confirmed as being from the *White Bird*, then the ocean grave of the two airmen would be just off the coast of Ireland, instead of thousands of miles away off another continent.

Such events were in people's minds at that time, for in the first fortnight of September 1927 no less than three transatlantic attempts had failed without a single survivor to tell the reason why. One of these was a westbound flight by a single-motor Fokker crewed by two experienced ex-RAF war pilots, Lieutenant-Colonel Minchin and Captain Leslie Hamilton, with the sixty-year old Princess Lowenstein-Wertheim as passenger.

The princess was English by birth and her ambition was to be the first woman to fly the Atlantic. She certainly was the first to attempt it, and she may even have succeeded. In mid-Atlantic the Fokker *St Raphael* passed over the Standard Oil tanker *Josiah Macy*, and exchanged signals by Aldis lamp. She had covered 900 miles, and headwinds had reduced her speed from 110 mph to 85 mph, but she still had plenty of fuel to make Newfoundland. But once again, as it had during the expected time of arrival of Nungesser and the *White Bird*, fog blanketed Newfoundland and much of the offshore ocean, and the wind went round to southwest. The Fokker may have gone down in the sea, but it is more likely that it went off course to the north over an unexplored wilderness where a crashed plane might never be found, or not for a century.

The unknown fate of the westbound Fokker did not deter two other teams then preparing to fly the Atlantic eastwards. *Old Glory* took off for Rome, flown by Lloyd Bertaud and J.D. Hill, with Philip Payne as passenger; 600 miles out over the Atlantic from Maine, they sent an SOS. Ships headed for the position given, but all they found, four days later, was wreckage; there were no bodies, and no clue as to what had gone wrong. Even before the wreckage had been found, another plane took off from Newfoundland. The *Sir John Carling* was flown by two British pilots, Captain T.B. Tully and Lieutenant J.V. Medcalf; they were never seen again, and left no message, not even an SOS.

Even quite narrow stretches of sea could be lethal, could swallow without trace. Two of the best of the early British aviators disappeared in the narrowest part of the English Channel. In 1910 Cecil Grace, flying a Short-Farman capable of 60 mph, was in a hurry to get back to Dover from the French coast, although there was fog. He arranged with the captain of a ship, the SS *Pas de Calais*, to take off just after she left and follow her funnel smoke as a guide to direction. But the ship was late in leaving, and Grace disappeared into the fog alone. There were many witnesses in the crowded Straits. The Farman was heard as it passed over the North Goodwins Lightship, and a fishing boat sighted the plane near the East Goodwins. But Cecil never arrived at Dover. His cap and goggles were picked up on a Belgian beach two weeks later. His aeroplane may have been swallowed by the sands that have devoured thousands of ships over the centuries. In which case, it is still there, and capable of discovery.

Gustav Hamel went down in the English Channel on 23 May 1914 in a special racing Morane-Saulnier monoplane which he had just collected from Paris. He was warned that the 80 hp Gnome rotary engine had been giving trouble, but it did not let him down on the two overland 'hops' which took him from Paris to the coast. He refuelled for the last time at Hardelot and then took off for Hendon. No one saw him crash, and although the British Admiralty sent a flotilla of destroyers to scour the Straits of Dover for two days, they found nothing, not a trace. England lost not merely a national hero of the time, but probably also the honour of making the first attempt at an Atlantic crossing. A Martin & Handyside monoplane, with a 215 hp Sunbeam engine, was being built for him specially to make the attempt.

The year after Nungesser vanished from the sky, the French provided a plane and pilot for a Norwegian hero, Roald Amundsen, the Polar explorer who had come out of retirement to lead the search for the airship *Italia*, missing somewhere between Spitsbergen and the Pole with sixteen men. Amundsen, one of the greatest explorers of all time, was a proud, prickly ambitious man who had so embarrassed his government that they would not support him and he had to find backing in France. Just before he left Oslo, he told an Italian journalist of his feelings for the Arctic: 'Ah! If you only knew how splendid it is up there! That's where I want to die. I wish only that death will come to me chivalrously, will overtake me in the fulfilment of a high mission, quickly, without suffering.'

It was not to be.

On 18 June 1928, Amundsen was at Tromsö in north Norway with a prototype flying boat, a Latham 47 twin-engined biplane, which had been flown up from Caudebec-en-Caux by its French crew. The pilots were Commandant René Guilbaud and Lieutenant Albert de Cuverville, with Emile Valette as wireless operator and Gilbert Brazy as mechanic. Amundsen brought with him a third pilot, his Norwegian compatriot Lieutenant Leif Dietrichson, who had flown a Dornier Wal for him in 1925. It was a quarter of a century, almost to the day, since Amundsen had left Norway in 1903 on that Polar expedition which had resulted in the discovery of the Northwest Passage, sought for centuries by seamen of many nations. Now he was just a retired Polar explorer and this was his last chance for the limelight. The fact that the missing Italian airship was commanded by General Umberto Nobile, for whom Amundsen had conceived a bitter rivalry after their joint Polar flight in the *Norge* back in 1926, gave additional savour to the situation. But Amundsen sealed his own fate. A number of large rescue aircraft were gathering in the north, including a Swedish tri-motor floatplane piloted by Viktor Nilsson and an Italian Savoia S-55 twin-motor flying boat flown by Major Maddalena, the Italian ace.

Nilsson suggested that they go on to Spitsbergen in company, so that if one machine was forced down the other could report or perhaps rescue the crew. Amundsen gave him no reply, because, with that competitiveness which had taken him to the South Pole before Scott, he conceived of this search and rescue operation as some sort of race, a test of personal and national prowess. He, Amundsen, for Norway; Nilsson for Sweden; Maddalena for Mussolini, whom Amundsen detested. When he heard that Maddalena had taken off from Vadsö, further north, Amundsen at once ordered the crew of the Latham to get airborne and, without waiting for Nilsson, the French flying boat roared across the water and took off for Spitsbergen. And was never seen again. No message ever came from it, but that was not surprising, as it carried a weak wireless set having a range of about 100 miles only when airborne and which would not work at all when the Latham was on the water.

Scott's faulty planning had led four men besides himself to their deaths in the Antarctic in 1910. Now Amundsen's ill-considered rashness in old age had killed five men besides himself. But not in the Arctic icescape, as he had wished, and certainly not quickly and mercifully.

No written message over came back from the missing Latham, and none of the aircraft diverted from the search for Nobile to look for it found any trace of Norway's lost Polar hero. The French machine vanished on 18 June and it was not until 31 August that the little ship *Brodd* sighted a large floating object off Tromsö. It rode high in the water, like a buoy or a barrel; but it was neither. The find was of very light metal, painted blue-grey and about seven feet in length, just like the wingtip float of a flying boat, and much battered. When it was shown to a mechanic who had worked on the Latham at Bergen, he discovered a convincing detail, a small copper plate welded over a tiny puncture in the float in order to keep it watertight. He had done that repair himself and recalled it well. The now battered state of the float was evidence that the Latham had hit the water with great violence, wrenching it loose. That, it might be thought, indicated that Amundsen had met the quick end which he desired, but a later find showed that this was not so. Again, there was a small, convincing detail as eloquent as any written message.

This time the container was a petrol tank, also found floating, also showing signs of what at first sight appeared to be a repair. But whereas the wingtip float had been properly welded tight in a workshop, this was a petrol tank with a hole filled by a wooden plug rough-cut to fit with something like a pocket knife. It was identified as the lowest fuel tank in the Latham, the one it would be easiest to get at and remove. Why should they do that? Well, it was known that the French co-pilot, de Cuverville, had once reported favourably on a subordinate who had lost a wingtip float when forced down on the sea, but had got off again by doctoring an empty fuel tank so that it could be fitted in place of the missing float. The theoretical pieces now fitted neatly. The co-pilot knew of the method, one wingtip float had indeed been torn off (the first find) and here the second find was of a doctored fuel tank. Amundsen and his companions had not died quickly and mercifully in the frozen north but in a final battle for life in the waters of the Gulf Stream.

The evidence from planes known to have come down in the sea was not always so clear. An instance was the disappearance on 8 November 1935 of Sir Charles Kingsford Smith, somewhere off the coast of Burma. Ira Jones, who was so scathing about Richthofen, regarded 'Smithy' as the world's greatest long-distance flier. An Australian, 'Smithy' joined up in 1915 on his eighteenth birthday. He flew Spads with No. 23 squadron, won the Military Cross, and was wounded. He was still limping from that wound when Jones first met him in France. After the war, anxious to continue flying, he got into the hand-to-mouth business of giving joyrides and air-taxi work. A prize of £10,000 was offered for the first flight by an Australian from England to Australia, but the prime minister of Australia in person banned 'Smithy' from undertaking it on the grounds of his youth and inexperience.

This officially unpromising and untalented young man subsequently made a succession of brilliant flights, including a round-the-world trip and the first crossings of both the Pacific Ocean and the Tasman Sea. But there was a long, hard road of professional flying and, most difficult of all, fund-raising involved. He did stunt flying and wing-walking for Hollywood films and at barnstorming meets, but gave it up, he said, because American audiences were too bloodthirsty. They didn't just want to see a man walk on a wing, they wanted to see him fall off it, too, and the body carted away. This was a keynote of 'Smithy's' career: he worked everything out so as to eliminate chance as much as possible. He took risks, of course, but they were meticulously calculated. A passenger would feel safe with him. Indeed, even after he had become famous and had been knighted, he had to earn his bread by giving joyrides from dawn to dusk in the old Fokker tri-motor *Southern Cross* with which he is most associated. Some people felt that this was demeaning, now that plain 'Smithy' had become 'Sir Charles', but the man himself was unable to grasp how earning one's living honestly could be a matter for scorn.

Jones met him for the last time the day before he left Croydon in 1935 in a fast, new machine with the object of beating the record set up by the winners of the recent England to Australia air race, Charles Scott and Campbell Black. 'But he was not his usual cheery self,' wrote Jones, 'he was bordering on morbidness. He looked worried and I ventured to suggest that he was not looking well. "Perhaps not," was his laconic comment.'

'Smithy' may have been recalling the fate of his best friend and colleague, C.T. Ulm, who had disappeared on a Pacific flight the previous year. Flying an Airspeed Envoy, built at Portsmouth, he and his crew are believed to have gone down near Honolulu. Another colleague of Australian flying who outlasted them all was Jim Mollison. Australia was a rough continent for aviation then. A forced landing might be in a sun-scorched heat-blasted wilderness which would kill far more quickly than the pack-ice ot the Arctic. But nothing was, or is, quite so swift to kill as water.

The *Southern Cross* had had the safety factor of three engines – 220 hp Wright Whirlwinds. It was a slow, clumsy machine with a top speed of 120 mph and a cruising speed of 94 mph, but it could maintain height on two engines and fly a long way on one. His new machine, the *Lady Southern Cross*, was a Lockheed Altair – a fast, low-wing monoplane with retractable undercarriage powered by a single Pratt and Whitney Wasp engine. 'Smithy' had wanted to use a De Havilland Comet, a twin-engined monoplane with a top speed of over 200 mph and a range of 2800 miles, which was the only British aeroplane which could be considered. However, he could not obtain one with a variable-pitch propeller in time, and instead chose the Altair with its very similar performance on a single engine, which was in standard production. It was a beautifully clean aeroplane – no struts, wires or stuck-on gadgets to impede the airflow – and with a variable-pitch propeller had a short take-off run and comparatively low landing speed, coming in at 85 mph and touching down at 55.

Kingsford Smith left England on 6 November 1935 with Tommy Petheybridge as mechanic. He disappeared at night off the coast of Malaya on 8 November. Up to that time, he was flying ahead of the record he had set out to beat. His line of flight on 7 November had been Allahabad to Calcutta in India, then Akyab to Rangoon in Burma, and then the same course continued out over the Gulf of Martaban. How far along that line did the *Lady Southern Cross* travel before whatever happened, happened?

Some two years later, in 1937, a wheel from the Altair was found floating in the sea off Aye Island in the Gulf of Martaban. An Australian engineer, Jack Hodder, visited the island and was convinced that he had found traces of an aircraft having struck tree-tops and broken them, leaving some pieces of metal behind, to give a guide to the final flight path of the plane, which must have dived into the sea somewhere offshore. The island has a then uncharted peak which might well trap a low-flying aircraft whose pilot was not aware of the hazard, and it was on this peak that the trees were broken. In May 1979 this finding was followed up by E.P. Wixted of the Queensland Museum, who was sufficiently convinced by his own visit to start planning an underwater search off the island for the remains of Smithy's Altair.

Quite a different suggestion was made by Sir Lawrence Wackett in his Kingsford Smith Memorial Lecture in Sidney, Australia, in September 1974. Sir Lawrence possessed information which made him feel certain that the Wasp engine had failed, and failed for a particular reason. It was an 'E' model of the Wasp series, an experimental engine in the development of the series. He himself had selected the Wasp when the Commonwealth Aircraft Corporation was set up to build aircraft and engines in Australia, but by that time Pratt and Whitney had reached the 'H' model Wasp. He had asked about the new features they had introduced and 'particularly noted that the speed of the supercharger impeller had been reduced from 12:1 to 8:1 so that the rated altitude for the "H" model was now 6000 feet instead of 8000 feet rating for the "E" model. It had been proved by exhaustive testing that the bearing of the impeller had a limited life at the higher rpm and some failures had occurred on test engines which had been run to destruction.'

At the time of 'Smithy's' flight, no one would have known that. Kingsford Smith himself may have driven the engine hard in order to break the record, not suspecting that there was a weakness which previous flights had not revealed. Moreover, it was a weakness which must result in complete failure of the engine.

For the one reason or the other, the *Lady Southern Cross* must have gone down somewhere in the Andaman Sea, probably off the Burmese coast.

The Soviet equivalent of Knighthood is 'Hero of the Soviet Union', the honour borne by Sigismund Levanevsky when he disappeared in a four-motor passenger machine during a proving flight across the Arctic Sea from Moscow to Fairbanks in Alaska via the North Pole. The machine had a picked crew of six and three were Arctic veterans.

Levanevsky had had an incredible career, largely due to the Revolution. The son of a blacksmith, he had to work from the age of twelve to support his widowed mother. In 1917 he was an unskilled factory labourer who joined the Red Guards. Here he found his metier. At seventeen he was commanding a battalion, and rose to be second-in-command of a rifle regiment. After the civil war he became involved with military balloons and then with flying boats and seaplanes. Seated at the controls of a four-engined H-209 he took off from Moscow on 12 August 1937 and all went well during the early, critical stage of the flight when the heavy load of fuel prevented the machine from climbing. At this time of the year, ice sometimes forms inside clouds but for the first 2000 miles Levanevsky was able to avoid this hazard.

By the time they did meet dangerous cloud, the machine was light enough to climb up into the bright sunshine and dry air at 20,000 feet; and at that height, with nothing to be seen below them but the billowy glittering cloudtops stretching to the horizon like hills of unearthly snow, they met their first setback – an adverse wind of about 62 mph which was cutting their speed by nearly the same amount. They were lumbering along at the pace of a motorcar, an eternity of time and boredom between them and their goal, which lay deep inside Alaska behind a high mountain range. That was thousands of miles away.

With the North Pole two hours behind them, on the downhill run to the New World, they transmitted a distress message – the only one which, for certain, was received from the Soviet plane. One of their four engines had stopped because of damage to an oil-line; the loss of power had forced them down to 13,000 feet and at that level, inside the grey, freezing clouds, ice was forming on the aircraft. The last words most listeners heard were: 'Do you hear me?' A few other listeners picked up half of another sentence: 'We are landing in . . .' After that, there was only a jumble.

The assumption on that evidence was that the pilot had decided to make a precautionary landing – as opposed to a 'forced' landing – in the hope of either repairing the engine or waiting for better conditions, or both. On the other hand the summer, the time of thaw in the Arctic, would tend to break up the few level surfaces there normally are on the pack-ice. A crash might result, or even be accepted, and then the survivors could either wait for rescue or attempt to walk out to safety. A great deal might depend upon whether or not there was a portable radio on board which survived the crash or crash-landing. A wireless powered by the engines would be of limited use.

In 1928, a portable radio had survived the crash of the *Italia* on the pack-ice and after many days of fruitless transmissions a message was picked up. But the real messages had to compete with reports which were either totally imaginary or deliberately false. Eventually, the matter was settled by asking the man who claimed to be Biagi of the *Italia* for Biagi's military number: when it was found that he knew it, that clinched the matter, and the search could be launched in the right direction instead of in several wrong areas. However, it did not follow that all radio 'hams' were well-meaning wishful thinkers or outright hoaxers and that official, government wireless stations were beyond criticism. Far from it. The faint but genuine SOS messages were not heard at first by the Italian government base ship at Spitsbergen, largely because they had given up the airship crew for dead and were transmitting news instead of listening for it. The first message to be picked up was by a Russian amateur wireless man more than a thousand miles further away. The survivors in the event owed their lives to him and not to their colleagues whose duty it was to listen for them.

After the last jumbled words from the H-209 on 13 August, a number of messages which seemed to be from the Russian plane were picked up. Some might perhaps have been genuine, but the listeners did not manage to decipher a position from what little they heard. Over a period of some days the calls became weaker and then stopped, as if batteries had become exhausted. But where were the searchers to look? All they had was a straight line running from the North Pole to Fairbanks in Alaska. Air searches were flown by Americans, Canadians, Russians. Nothing was seen, but that did not mean that nothing was there. The experiences of the *Italia* survivors showed that the jumbled nature of the pack-ice created a multitude of false outlines in which a complete aeroplane was impossible to see, when in addition the human eye was working in the conditions of super-brilliant, excessive sunlight and blinding reflection. Earth, woods and water absorb a good deal of sunlight; ice does not. Time and again, aeroplanes were seen by the survivors, but the crews of the planes did not see the 'red tent' on the pack. The first aeroplane actually to pick up the tent lost it again immediately afterwards, when the pilot looked away momentarily, and he had to be directed back by Biagi's portable radio down below.

The Italians had been some seven weeks on the pack when they were picked up by a Russian icebreaker. There had been instances where men from ships crushed by ice had survived for one, two or three years. This was normally when they could live by hunting, shooting, or fishing for fresh food. The Soviet flyers had supplies for three months, plus rifles and ammunition; they were well-equipped for survival. If they lasted into their first winter, the search planes might stand a better chance in the Arctic moonlight of seeing them than under the intense glare of summer daylight. Searching went on for a full year, but nothing was ever found.

The common assumption was that they had come down two or three hundred miles from the Pole on the Alaska side. That was not a fixed position, because it was now known that the pack-ice drifted under the influence of both wind and current. The prediction of the Soviet authorities, based on the drift of the marooned Papanin group earlier, was that within a year the ice-floe they were on would reach the Gulf Stream and melt underneath them. If so, then on 12 August 1938 they would be dead and any search after that would be pointless. However, there were places where currents divided into two and others where great, swirling eddies occurred. Iced-up ships had been known to go round in circles, frozen into the pack. So there was just a chance they might have drifted to Greenland or Alaska.

There was also a more sombre solution, with some evidence to support it. If the Russian plane had come down through the clouds safely, without severe icing, and then lumbered on powered by the three good engines, the next point of danger for a low-flying machine was the Brook range of mountains in northern Alaska. This area was searched and the Eskimos questioned, or rather, gossiped with, because that was best calculated to produce remarks unprompted by a questioner. On Barter Island only, not far off the true course Levanevsky intended to fly, the Eskimos had heard on 13 August what at first they took for a power-boat and then realised must be an aeroplane, coming from seaward hidden in clouds, and continuing inland.

That was interesting evidence but hard to check. If the machine, large though it was, had flown into a mountainside, it was unlikely to be found. There would be no survivors, snow would soon cover it, and anyway the very few foot travellers in those parts kept to the trails. They were not likely to go wandering off into the wilderness. Perhaps some day, if the plane is there, during a thaw some prospector for gold or lone hunter might come across the wreckage of that antique machine and the six Russians who flew it. But it is not very likely.

The Soviet heroes are forgotten, but the mystery of Amelia Earhart's disappearance remains. The headline career of the American heroine began in 1928. The same newspapers which were carrying the news of the search for survivors of the lost airship *Italia* simultaneously used banner headlines to tell of the first woman to fly the Atlantic. This was unfair to the pilots of the Fokker tri-motor, for Amelia was along only as a passenger. But when she vanished in the Pacific in July 1937, a month before the Russians were lost over the Arctic, 'AE' had been sitting in the pilot's seat of record-breaking aeroplanes for a long time. Her disappearance, however, may not have been what it seemed. This suspicion of an inside story yet to be told is what keeps the mystery as evergreen as that of the *Marie Celeste*, and there is always someone to undertake a new research project into the fate of Amelia Earhart.

The first hint came within a few months of AE's disappearance and a few weeks after the Russians had gone down. In the September 1937 issue of the American magazine *Popular Aviation*, Swanee Taylor wrote: 'If nothing else (and there was plenty if you know the inside story), Amelia gave our Navy a priceless opportunity to demonstrate to the belligerent world that this country is well equipped to take care of itself in the Pacific.' What was that inside story? Did AE and her Irish navigator Fred Noonan die on US government business? Or were they merely unlucky at the riskiest part of a daring flight? It was planned in fact to cover a politically sensitive area, the scene of many island-hopping battles to come in the Second World War, which many Americans and Japanese expected.

Indeed, with the year 1937 we are into a new war period with new personnel. The old fliers of the Great War were now either dead, retired, or operating at a high level of organisation and command. The new names were mostly people who had been too young for the 1914–1918 conflict. Had she been born a boy on 24 July 1898, Amelia might just have made it. As it was, her flying lessons had to be spaced out, because they had to be paid for. Amelia graduated from Columbia University in 1919, took lessons in 1920, soloed in 1921. Flying was what she really wanted to do, but she had to take a job as a social worker. Drink cases, she found, made up half the family problems she encountered. This was familiar. Amelia's mother had been the spoilt daughter of a rich man, but she had married someone without the same earning potential. The poor chap could slave his guts out, but the dollar result would be considered contemptible by his wife's father. Eventually, he took to alcohol and the marriage broke up. Amelia was desperately sorry for her father and was able to help him, and through this experience perhaps also the people she encountered in her social work. Certainly she considered also that a major factor in many divorces was the excessive dependence of the female on the male, so that when the latter met hard times, the wife was unable to play a helpful part.

In 1927, the man she was eventually to marry, the publisher George Palmer Putnam, had succeeded in persuading Lindbergh to sign a contract to write a book about his Atlantic solo flight, which had caught the imagination not merely of America but of the whole world. Now he was thinking of some woman author to write the story of the first woman to fly the Atlantic, but as no woman had done the trip yet, the would-be author would have to do that little thing first. His first choice for an Atlantic aviatrix who would turn author for him, had second thoughts and backed down, although he had a plane and two men ready to go. All George Putnam needed was a modest lady who would accept Putnam's modest money, possessed a current pilot's licence, was brave, intelligent and had some writing talent. Not a great deal to ask, but he got more. Amelia Earhart, who was slim and wore her hair cut short, bore a striking resemblance to a female version of Lindbergh. So he had a 'Lady Lindy'. She had a further very necessary quality which he had not suspected. The pilot of the Atlantic Fokker was to be Wilmer L. Stultz, and he had an alcohol problem just as her father had. During the long, nervous wait on the North American side of the Atlantic, Amelia's expertise in this line was very much required, and on 17 June 1928 Stultz made a brilliant flight from Newfoundland to Wales. Putnam got his next bestseller and 'Lady Lindy' was famous. An honest person through and through, she herself was not really happy with the applause until in 1932, already married to Putnam, she flew the Atlantic solo in a Lockheed Vega, Newfoundland to Northern Ireland.

By 1936 virtually only one worthwhile flight remained for AE to do – round the world. She was given a Lockheed Electra, a superlative machine for its time, a small, twin-engine ten-seater airliner which she could convert into a comfortable long-range plane. Its endurance in still air was calculated at 4500 miles, but the Pacific is broader than that. The US government, entirely at its own expense, built no less than three airstrips on tiny Howland Island, turning it into an intermediate staging post for AE's world flight. They could always claim that this was an entirely civilian concept and that no thought of Howland's future potential use as a strategic airbase had so much as crossed their diplomatic minds.

In March 1937, starting from California, she reached Honolulu; and there she crashed on take-off. The plane, for which Pardue University had paid 50,000 dollars, had to go back to the factory for repairs. Of her two navigators, one had used up all his leave. When AE took off again for the round-the-globe trip it was to fly in the opposite direction, starting with the Atlantic instead of the Pacific, and with only the one navigator, admittedly one of the world's very best, Fred Noonan. Basically he was a seaman and had survived being torpedoed three times during the war; his forte was astro-navigation – that is, the use of sextants to fix the position of sun, moon and stars. Over trackless oceans with few landmarks or none, this was a vital skill.

Unfortunately, Fred could empty a bottle of whisky in an afternoon and still have room for more in the evening. AE seemed to collect them. So far it had not affected his functioning in flight, only on the road. Noonan had just got married, and he celebrated by driving into another car head-on. Alas, he was on the wrong side of the road and had taken more than one too many. However, Amelia believed in him and the flight went like clockwork right up to the last, except for one navigational error over Africa, which was due to AE overruling Noonan's reckoning (which was correct) and flying the course she thought was right (which in fact was wrong).

The only person not quite happy with Fred Noonan was George Putnam, her husband. He had made Noonan sign a contract, which AE had voided, in which Noonan promised to stay out of the publicity limelight and let Amelia have it all. From the publishing point of view, doubtless this was good business, but it was not entirely all. Putnam made the money (e.g., 25,000 dollars for letter-covers carried in the Electra), but he hungered for a piece of his wife's publicity. At press conferences he was always edging into camera and got himself the nickname of the 'lens louse'. Amelia, who believed that wives should in adversity be able to assist their husbands, had gone rather too far in this direction for George's liking: she had eclipsed a successful man, in the media at any rate.

On 2 July 1937, Amelia had almost completed her round-the-world flight. On reaching Australia she and Noonan had left their parachutes behind. Over the Pacific, these would be so much useless lumber, and most of the 7000 miles remaining were water. From Lae in New Guinea her next hop was to Howland Island, 2556 miles away. There were no landmarks, no checks on the navigation except by the sun, moon and stars – if they could be seen. And Howland Island was two miles long by half a mile wide and nowhere more than twenty feet high. A speck in the immensity of the Pacific Ocean.

In normal conditions, such a flight would be a mad chance. However, without consulting the taxpayers, and for what on the surface was a purely civilian flight, the US government was providing support: the USS *Ontario* was positioned halfway between Lae and Howland, and actually at Howland was the Coast Guard cutter *Itasca* with homing and direction-finding equipment. Additionally, the USS *Swan* was stationed on the other side of Howland, halfway to Honolulu. The resources involved, which included the gift of a modern, twin-engined airliner, the loan of three warships with the latest secret radio equipment, and the construction of three airstrips on a Pacific Island, were wildly beyond what any ordinary flyers could expect. Even Kingsford Smith never had anything like this, and most of the long-distance pilots set off in light aeroplanes with a range of a few hundred miles and a cruising speed of 80 mph.

The ground station at Lae kept receiving messages from the Electra for almost half the distance to Howland Island, some 1200 miles, where AE was due to arrive next morning after an eighteen-hour flight. But off Howland the *Itasca* found very great difficulty in picking up Earhart's signals. There was a lot of static, her voice would fade, or become unintelligible. They never heard her for long enough to get a satisfactory bearing. And apparently she only heard them once, at 08.03 when matters had become critical. At 02.45, AE said something about 'cloudy weather cloudy . . .' At 03.45 a sentence came through, with Earhart asking the *Itasca* to broadcast on 3105 kilocycles each hour and half-hour. At 04.53, just snatches of Earhart voice, nothing understandable. At 05.12 she could be heard asking for bearings to be given her on 3105 kilocycles – she would whistle into microphone so that the *Itasca* could take a bearing. This was followed by a brief whistling, which did not last long enough for a bearing to be taken. At 05.45 a sentence came through: 'Please take bearing on us and report in half-hour. I will make noise in microphone. About 100 miles out.'

Then the static closed in and submerged everything for a long time. In addition to its ship direction-finder, the *Itasca* had set up ashore on Howland a high-frequency direction-finder which AE did not know about, but this was unable to secure bearings either.

Between 07.30 and 08.03 parts of three messages from the

Electra were heard. The first said: 'We must be on you but cannot see you but gas is running low. Have been unable to reach you by radio. We are flying at 1,000 feet . . .' The voice was calm, but the atmosphere in the cabin of the Electra must have been tense. At 07.57: 'We are circling but cannot see island. Cannot hear you . . .' At 08.03, for the first time AE said that she had heard a transmission from the *Itasca* but had been unable to get the direction of it, and asked again for the ship to take bearings on her signals, letting her have the results on 3105 kilocycles. She then began to transmit but the dashes faded quickly and then were lost in the static.

At 08.44 came the last message: 'We are on the line of position 157–337 . . . we are now running north and south.' Effectively, they were lost. It takes two lines to give a 'fix'. No bearings had been obtained either. So the Electra was flying on a north-south line somewhere. It could be to the north of Howland or to the south. The plane could have overshot Howland to the east or undershot to the west. It could be near or it could be far. But there was no more word from AE, at least not certainly.

One of the most expensive searches in history was launched, including an aircraft carrier from San Diego and a battleship with three floatplanes from Pearl Harbor. Various garbled distress messages were picked up, one or two of which just possibly might have been genuine; then a storm swept the sea. The distress calls continued but hope did not. If the Electra had been floating, it must by now have sunk; if AE and Noonan had got ashore on a coral reef, the waves would probably have swept clean over it. No trace of the Lockheed and its crew was seen; none have been found since. AE and Noonan vanished utterly without trace.

In 1942 Hollywood made a movie out of the story, thinly disguised, with Rosalind Russell as Amelia Earhart, Fred Macmurray as Fred Noonan, and Lockheed Electra NR16055 as Lockheed Electra NR16020. Its title was *Flight for Freedom*, and the plot anticipated Gary Powers and the U-2 incident by many years. Some people quite reasonably had suspected a straight spy mission: the Lockheed would 'accidentally' go off course and just by chance fly over the Japanese Mandated Territories in the Pacific which the American government were very curious about. There very well could be excellent cameras in what AE used to call her 'flying laboratory'. The Japanese were not supposed to fortify

Last Flight of Amelia Earhart

these islands any more than the United States was allowed to put guns on Howland, but Howland now had three airstrips. What, for instance, did the Marshall Islands have in the way of newly-built quays and aerodromes? This type of information is much more important than a few guns, because it tells an opponent if the area can or cannot in the near future be used as a launching ramp for a substantial offensive.

The movie started with this premise but had to complicate matters for the sake of 'plot'. Rosalind Russell has to fly to 'Gull' Island and pretend to be lost over the Pacific, which triggers the 'search' for her by US Navy planes during which some will 'accidentally' stray over Japanese territories. But before she takes off from Lae in New Guinea Rosalind Russell learns that her mission is 'blown' and the Japanese will land on 'Gull' Island and capture her, so she cannot send out distress messages. What does our heroine do? She deliberately crash-lands in the sea, so that the US Navy search can go through, as scheduled.

A long time after the war, with the trail very cold, an American radio reporter called Fred Goerner, who was also an amateur pilot, spent years researching stories of two white fliers who had been captured and executed by the Japanese before the Second World War. The reports concerned an injured man and a woman who was tall, thin and wore her hair cropped short like a boy. Goerner investigated the sunken wreckage of a plane (which turned out to be Japanese), dug a grave site (which contained the remains of four or more orientals), made a nuisance of himself to the CIA (who were running a hush-hush spy establishment he stumbled over in his quest), but did obtain a disturbing number of native witnesses who told him about the two American fliers who had crashed in the Marshall Islands and later been taken to Japanese headquarters at Saipan for questioning. The man had been executed by decapitation, the woman had died in prison of dysentery. There was nothing unbelievable about any of that. The convention was, and still is, that agents work on the understanding that, if discovered and captured, they will be disowned by their own government so as not to give provocation for war.

There are much more unlikely stories around – for instance, a book in which the author claimed to have met Amelia Earhart in 1965. There are only two broad alternatives: that AE and Noonan perished in the Pacific a hundred miles or so from Howland Island as a result of navigation problems, or that they died for their country at the hands of the Kempetei in Saipan. For their sakes, one hopes it was the former.

'If I were asked whom I considered to be the three greatest women pilots I should unhesitatingly say Amy Johnson, Jean Batten, and Amelia Earhart, with Amy the greatest of all; it was she who blazed the trail,' wrote Squadron Leader Ira Jones. Even discounting something for the Squadron Leader's patriotism, he had a point. But Amy Johnson's undoubted popular success with the British public owed a good deal to a legend which is current to this day. In the spring of 1980, when there were celebrations to commemorate Amy's solo flight from England to Australia in a DH Moth with a 100 hp engine, a journalist could still recall 'her previous career as a typist or shopgirl'. The image of the young girl who came from behind the counter to fly an aeroplane all the way to Australia in primitive conditions was irresistible. But it is not quite true. Amy's background was very similar to Amelia Earhart's. She too was a university graduate, and although not vulgarly rich, her people were comfortably off. You had to be, to fly privately in those days. An hour's flying lesson could cost £2. 10s., a good weekly wage in the 1920s and an hour's solo 30/-, even at government-subsidised flying clubs.

Only the very rich flew their own aeroplanes. The rest saved up, or sponged on their parents, to buy one half-hour flying lesson a week, or a fortnight, or a month. That was how I learned in 1933 in a Cirrus II Moth identical to the one Amy had learned on in 1928. But I was a schoolboy of fifteen then, and no one resented me. Some were to resent Amy greatly.

Her first lesson and the instructor's comments rather recall the time when the future Captain Albert Ball, VC, was advised to find a flying school for girls and join it. Her aptitude was obscured by three facts: she was a small girl and could hardly see out of the cockpit; the helmet she had borrowed was loose-fitting and she could hardly hear what the instructor was saying through the earphones; and the instructor was unsympathetic and brusque. Anyone of less determined character might have packed it in there and then. But she did carry on with her lessons and she did pay for them out of her salary first as a shopgirl and then as a typist, rather better-paid jobs than she would have had had she gone on to be a teacher. Perhaps part of her determination was because she had recently been jilted after a long love affair.

Learning to fly was one thing, flying to Australia quite another. Ground facilities along the route were either unknown, meagre or non-existent, or so one gathered. There was likely to be trouble with passports, visas or innoculations, and natives who did not speak English. And if you were going to do the trip fast, you virtually had to do without sleep for a week, and eat foreign food likely to give you 'gippy tummy'. This was where the experienced pilot, who could virtually fly in his sleep even while suffering from malaria, had the advantage. And then there was the weather. Flying light aeroplanes in England was a fair-weather sport; but if you went to Australia in one you met half the world's weather, some of it bad. The first such flight had been made as long ago as 1919 by a converted twin-motor Vickers Vimy bomber with a professional crew of four. In 1928 Squadron Leader Bert Hinkler had caused a sensation by flying there solo in a light aeroplane in sixteen days. There would be headlines now only if a girl repeated the performance. A man was no good, even if he was as refreshingly amateur as Amy. Just such a character, Francis Chichester, attempted the flight but aroused no interest. He had to wait until almost the end of his life, and then go round the world in a yacht called *Gipsy Moth*, before gaining the applause; and even then, it is hard to grasp why his feat in media terms so far eclipsed that of Alec Rose in *Lively Lady*.

While she was struggling to obtain support for her venture, Amy spent much of her spare time studying for her commercial pilot's licence, aero engineering and airframe licences and, most difficult of all, a navigator's licence. Eventually, she obtained the aid of Sir Sefton Brancker, Director of Civil Aviation, who was impressed. 'I felt that although the odds were against her, she was just the type to triumph through,' he told Ira Jones. Brancker thought that if an ordinary-seeming girl and only an amateur pilot made such a flight, this would impress on public opinion the safety of modern flying and combat the still prevalent image of the aviator as reckless daredevil, which was not good for business. He spoke to Lord Wakefield, who put up half the money to buy her a plane, and Amy's father then paid the other half. She had her machine just two weeks before take-off.

Professional pilots who did not understand that Amy's credentials were precisely her girlhood and inexperience became bitterly jealous when they saw her lauded by the press for feats most of them could have surpassed while hardly trying. Ira Jones was to hear some of them betting that she would not get beyond Vienna when she set off for Australia in May 1930. The Moth she was flying was second-hand and had the registration letters G-AAAH (the one I learned on three years later was G-AAAG, the machine immediately preceding it). Nevertheless she reached Karachi in India one day ahead of Squadron Leader Hinkler's time, and people began to take notice.

The future Australian record-breaker, C.W.A. Scott, was chosen to meet her in an aeroplane at Port Darwin and then escort her by stages to Brisbane. Amy had flown from a rather damp, coldish country across several continents southward to a boiling hot Australian afternoon. 'Imagine a sunburnt girl, wearing oil-stained shorts and shirt, the centre of an admiring and worshipping mass of people, each trying to shake her hand and wanting to carry her shoulder-high,' recalled Scott. 'There was no more modest person in the world than Amy Johnson at that moment. She was just a quiet, attractive English girl, of whom we were all justly proud.'

She had just crossed the Timor Sea (to which then the adjectives 'dreaded' or 'shark-infested' were normally applied) in a light aircraft without a mechanic to overhaul the single engine on which everything depended. Scott went over to look at it. 'Never have I seen an engine in such an appalling condition,' he commented. 'I for one would not have cared to cross 560 miles of open sea in that machine.' He and an engineer set to work to put it in order. Amy got lost before reaching Brisbane and finally crashed on landing there, incidents which could be recounted with relish by those who wanted to decry what the girl had done. Scott however thought that: 'Her whole flight was a wonderful performance. Allowances obviously had to be made at the end of such an ordeal. If she felt anything like I did later after a similar flight, she must have been in a highly overstrung condition.'

When Scott came to make his attempt in 1932, he made the trip in 8 days 20 hours, half Hinkler's time (Amy had taken three weeks). The following year Scott was leading an air circus round Britain which demonstrated the pulling power of female pilots. It had the Hon. Mrs Victor Bruce, Pauline Gower and Dorothy

Spicer, and a fascinatingly odd array of aeroplanes ranging from a 200 mph day bomber, the Fairey Fox, to a relic from the Great War, an Avro 504K with 130 hp Clerget rotary engine. When the circus came to the Portsmouth area I managed to get flights in both of these machines (father being in a good mood). This was a very optimistic time in aviation, for Britain held all three of the main world's records, for speed (407 mph), for distance non-stop (5341 miles), and for height (43,976 feet). This was the atmosphere in which Amy, still an amateur but with the public status of a record-breaking aviator, had to make a living in aviation. Eventually she married Jim Mollison, another record-breaking pilot, and they made a career together, not altogether a happy one. Mollison was a social character who spent a good deal of time at the bar and in the company of other women, a stylish and hectic life. Amy was a good, serious girl and this upset her.

However, the Mollisons made an Atlantic flight from Wales to near New York in a small airliner, a DH Dragon, and were both injured when Jim Mollison who was flying, ran out of petrol and crashed it. While convalescent they spent a weekend with Amelia Earhart and her husband and had lunch with the Roosevelts. The hard-drinking Mollison called the teetotal Amelia 'that strange, charming woman', while Amy admired her immensely. Possibly it was her fate that caused Amy a few years later to say that all long-distance flyers were bound to 'cop it' in the end. 'I know where I shall finish up – in the drink,' she told a friend. 'A few headlines in the newspapers and then they forget you.'

The Mollisons' marriage broke up, but Amy did not succeed in finding a regular job in aviation (as opposed to writing about flying for the newspapers) until June 1939, when she joined the Portsmouth, Southsea and Isle of Wight Aviation Company as a pilot. I recall seeing her there once, while I was waiting to take off, and she seemed happy and relaxed (it was common knowledge what a dispiriting time she had had with Jim, and that some professional pilots had given her a rough ride purely because she was a female competitor, during her vulnerable, amateur days).

Amy had over 2000 hours flying time in her logbooks when wartime reorganisation and anti-female bias combined to put her on the ground again. A lot of people in the RAF then felt that the RAF did not really require pilots. The German aces were to give them a shock shortly, but in the meantime the only way back into aviation for Amy was to join Air Transport Auxiliary (ATA), the ferry pilots organisation, whose nine-strong female section was headed by Pauline Gower. Even with the German Panzers pouring through Holland, Belgium and France the RAF did not require Amy for the army cooperation flying which she had been doing; indeed, the elementary reaction of everyone in authority in England at the outbreak of war had been automatically to close down everything – from the cinemas to the flying clubs – a policy of total paralysis, which persisted until the Stukas came.

Amy was thirty-seven when on 4 January 1941 she was ferrying a yellow-painted Oxford trainer, built by Airspeeds at Portsmouth, south from Scotland to Kidlington, near Oxford itself. The weather was bad but she managed to put down for the night at Blackpool, where her married sister Molly lived. She spent what was to be her last night on earth with them. Next morning, 'it was dreadful January weather, freezing fog. I told her she was mad to take off,' recalled Molly Jones, but Amy replied: 'I'll be all right.'

At the aerodrome, for a time visibility dropped right down, so Amy waited in the cockpit for it to clear, chatting to Harry Banks, a refueller. A message came from the Duty Pilot advising her not to take off, but she replied that she would fly over the top of the weather. The other two alternatives, flying in cloud or hedge-hopping under it, were both equally hazardous. However, going above cloud meant either finding a hole in the clouds near the destination or taking a chance and coming down blind through the murk until she saw the ground – which might not become visible until too late. Still, it was wartime, and people were expected to take risks.

At 11.49 that freezing morning the yellow twin-engined Oxford began its take-off run for the south. It had been a nasty, vicious day in the Thames Estuary, with high winds kicking up a swell and a confused sea breaking in the shallows over the banks. Intermittently, showers of snow brought visibility down to quarter of a mile or so. The water was only five weeks away from reaching its coldest point for the year. If held up by a lifejacket, an ordinary person might live for half an hour, a little more or a little less. At 3.30 that afternoon, with the light fading, the tidal stream was pouring through the deep channels that threaded the banks and connected the Straits of Dover with the Port of London, and yet another Channel convoy had rounded the North Foreland and was just entering the estuary. The small ships of the escort included HMS *Berkeley*, the AA trawler HMS *Haslemere*, which was towing a balloon, and the motor launch, ML 113. They had come from Portsmouth.

These convoys of little ships, mostly colliers and so known as the 'Coal-Scuttle Brigade', had suffered heavy losses in the all too recent past: as much as fifty per cent casualties had been inflicted on them by the Luftwaffe and the E-boats combined. They were crewed by civilians and almost always they were attacked during their passage of the Channel, most frequently in the Straits of Dover. But nowhere was really safe and the shapes of sunken ships sticking out of the shallow waters all along this coast like a grim paling fence of masts and funnels was a reminder of it. For them, the sight of dead bodies drifting by in the waves was no novelty. This was the front line of a war. Perhaps that was why a famous American newsman, Drew Middleton, was sailing with them.

The crews were still at action stations as they neared the East Knock John Buoy, but there was a lull. At that moment there were no guns thudding away or machineguns rapping angrily. Probably it was the noise of an aeroplane's engines unseen above the clouds which drew the lookouts' attention to a particular patch of sky 45 degrees on the starboard bow. The *Haslemere* was close enough to see a person on a parachute appear from the clouds in that direction and drift down towards the water. It was a not uncommon sight. Then an aeroplane came into view, about a mile away, and began to circle the parachute. Raymond Dean, a seaman, was sure it circled more than once; Lieutenant H.P. O'Dea thought it made three complete circles round the descending parachute. From HMS *Berkeley* and ML 113 only the aircraft was seen, not the parachute. But everyone saw the aircraft straighten out and go down into the sea at a shallow angle. Almost at once it began to break up and sink, leaving only floating wreckage.

The captain of HMS *Haslemere*, Lieutenant-Commander William Edmund Fletcher, had already altered course to starboard and ordered both engines to full speed ahead. When they were about half a mile from the scene, the parachutist hit the sea. The aeroplane crashed behind the parachute, further away from the ship. And then the trawler came to a shuddering stop, screws thrashing the water. She was aground, bumping heavily on the sands which stretch out many miles from shore. The tidal stream was bringing the parachutist almost directly down upon the ship, although the aeroplane wreckage would drift clear. The engines were put to slow astern and when the trawler slid off the sand, they were ordered to be stopped. By that time it was plain that there were two people in the water, not one. The furthest was about forty yards away, the nearest was brought by the current in towards the stern of the ship, which was pitching up and down in the confused sea. Both were wearing some sort of headgear.

Coming from off the bow the nearest person was carried in by the current very close alongside the after part of the *Haslemere.* Two heaving lines were thrown, one landing in front, one actually alongside, but the survivor made no attempt to grasp them or indeed move at all. Leading Seaman Nicholas William Roberts, who had thrown one of the lines, and was leaning over the bulwarks looking down directly at the person nearest, noticed the flying helmet worn and heard a thin shout of: 'Hurry, please, hurry.' 'It sounded like a small boy's voice at first,' he recalled, 'and then I realised it was a woman's voice.'

The woman had a second chance, apart from the heaving lines she had been unable to grasp, for Raymond Dean, an ordinary seaman, had gone over the side of the ship and down on to the rubbing strake, a projection which goes round the hull a few feet above the waterline. He lay down flat on this and with one arm holding on to the ship through the fairlead, he could reach down towards the water with the other towards the survivor who was only about five feet away. He too heard the thin call of: 'Hurry, please, hurry.' 'It was definitely a woman's voice and she looked young,' he testified. 'She appeared to have a life saving jacket on which was keeping her afloat. She did not appear to be swimming, there was no action of her body at all.' This was clear evidence of shock and exhaustion due to cold, the swift draining away of all feeling, all will. There was only one chance, for him to go into the water after her, carrying a line. Dean began to take off his heavy outer clothing, but was stopped by the captain, who told him to get back inboard.

Fletcher himself then took off his seaboots and dufflecoat and dived into the water. He was aged thirty-four and a very powerful swimmer, but in water of this temperature a matter of minutes only separated the strong from the weak. For some reason which will never be known, perhaps because he thought she was close enough to be safe, Fletcher did not after all swim to the woman. He made his way out towards the second person floating further away, often hidden by the rough seas as he did so. Dean saw this person also quite clearly, about fifteen yards away at least. He testified: 'I could not say whether it was a man or a woman or where it came from but I saw the features of a human being and I noticed that one arm was waving. This person appeared to be wearing a flying helmet, and there appeared to be a scarf or something muffled up round the neck.'

Leading Seaman Roberts testified similarly, but he was keeping a closer eye on the woman. 'She was now close in under the stern of the ship. The ship was heaving in the swell and the stern came up and dropped on top of the woman. She did not come into view again. I went on the other side of the ship, but the woman did not come into view. She had appeared quite calm and was not waving her arms.'

Lieutenant-Commander Fletcher reached the second survivor in the water and was seen to support him or her for a minute or so; then he gave up and made towards the *Haslemere*'s whaler which had previously been launched. Lieutenant O'Dea testified that this survivor had been floating upright, but when last seen was flat in the water. Lieutenant George Wright of ML 113 had the same impression, of a body floating face-downwards when last seen. Most of his attention, however, was devoted to rescuing Commander Fletcher, who was clearly in trouble now. The *Haslemere*'s whaler, crewed by seven men under oars, was making virtually no progress because of the wind and the waves.

Vivian Gray, one of the seamen in the whaler, recalled:

> We got very close to him and he shouted to me: 'For God's sake, lad, tell them to pull.' He appeared exhausted and just about done. One of my colleagues held on to my legs while I leaned over the bow and tried to reach Commander Fletcher with my hand but a big wave came and washed Commander Fletcher one way and the lifeboat the other. We made further attempts to get to him but did not succeed. When we were first drawing up close to Commander Fletcher I saw another person in the water which I took to be a German airman on account of the flying helmet that was worn. This person did not appear to be swimming but just floating and supported by something. I did not see any

> movement of the limbs neither did I hear this person shout . . . The head and shoulders were showing above the water.

In an endeavour to help Commander Fletcher, ML 113 put a Carley float over the side; then Lieutenant George Alexander Wright jumped over the side, got Fletcher onto the raft, and from there onto the ML. The captain of the *Haslemere* was then unconscious, and remained unconscious until he died shortly after. He had been in the water about twenty minutes, and the medical verdict was death by exposure and shock due to immersion. The *Haslemere* and ML 113 had been lying about 100 yards apart, with Commander Fletcher and the person he was trying to rescue between them. Meanwhile HMS *Berkeley* had sent her whaler out to investigate the floating wreckage of the aeroplane.

Lieutenant Ian David McLaughlan commanded this boat. With him was the ship's medical officer, Dr W.T.K. Cody. They found one wing of the aircraft broken off and floating; a short distance away two bags were adrift about ten yards apart. One bag was of pigskin and marked with the letters 'A.J.' in black. The other had the name 'Amy Johnson' written on it in ink. One of the bags contained Amy's logbook, so there was little doubt as to the identity of the woman flyer. Conclusive proof that the aeroplane had been hers was found the same day by Thomas Williamson, the skipper of HM Drifer *Young Jacob*, who was patrolling off Barrow Deep. He heard the news of the aeroplane crash and waited for the wreckage to drift down to him, recovering two pieces, the largest being ten feet long, of yellow-painted fabric on wood bearing black figures. He handed in the wood and kept a strip of the fabric. 'I then wrote on the fabric "*Amy Johnson Jan. 5th 1941*" and retained it for sentimental reasons as I am a native of Hull and knew Miss Johnson very well indeed.' Amy's father was a leading figure in the local fishing industry and was part Danish (his father's name was Anders Jorgensen).

A week later, Amy's chequebook inside a waterproof cover was washed ashore on the Isle of Grain, in the Estuary; and also a number of bodies. By then, however, Drew Middleton had written his account. He had first seen Amy's plane at 750 feet, gliding with engines off; then at 200 feet something white fluttered out of it; perhaps, but not certainly, a parachute. So the headlines followed . . . AMY JOHNSON BALES OUT, MISSING . . .

Then the Admiralty issued a statement saying that the crew of a trawler (the *Haslemere*) had seen two survivors in the sea – a man and a woman. They told how Fletcher had tried to save the man, but had died in the attempt. The only body actually recovered from the sea had been his. One of the two lost survivors was certainly Amy, but who was the other?

The speculation began. Most of it came from people totally ignorant of the conditions existing in the Thames Estuary which made it a 'hot spot' of the convoy wars, their focal point. This particular convoy had been under attack until just before the Amy incident. The doctor of HMS *Berkeley* had at one time counted eight parachutes in the air. He could still see some of them when the aeroplane came out of the clouds and crashed, because cloud base was fairly high at that moment. He thought, but was not sure, that the bombers were Italian. Mussolini's contribution to the air war against England was in fact based on the Belgian coast opposite and their targets usually were in the Thames Estuary. All sorts of strange-looking aeroplanes chose the convoys as their target.

When the London *Evening News* interviewed an official of ATA, the organisation to which Amy had belonged, he poured cold water over the whole Admiralty story of two people being seen in the water. Talking about the rough seas he had not experienced, he thought that: 'a natural mistake has been made, and the second object seen in the water was not a body.' The rival *Evening Standard* suggested that Amy might have had a passenger, perhaps some serviceman going on leave. It was Pauline Gower who poured the cold water on this one, saying that ATA pilots were forbidden to carry passengers without permission, and Amy was not the type to break this rule except for some life-or-death appeal. The Ministry of Aircraft Production were able to produce a great many witnesses from Blackpool aerodrome who could say definitely that she had been alone when she took off, but they gave the time of take-off as 10.45 instead of 11.49, stated that the 200-mile trip would take one hour and noted that Amy's plane crashed into the Thames Estuary (100 miles off course) at 3.30 pm. To be 100 miles out in a 200-mile flight is a remarkable discrepancy, as indeed it is to be airborne some three and three-quarter hours on a theoretical one-hour flight.

This spurred more speculation. WHERE DID AMY PICK UP MR X? asked the *Daily Express*. SHE WAS ALONE replied the

Daily Mail. 'Base, cruel and slanderous rumours' were denounced by the *Daily Sketch*. Supposedly, it was 'common talk' in the RAF that Amy had landed her Oxford (a twin-motor bomber-trainer) in some field to 'pick up a pal' and then for some sinister reason flown him out of the country to some other field on the continent. What she was doing over a heavily-defended area like the Thames Estuary in a large but low-performance aircraft was not explained. The German-held coast opposite was just as lethal in the circumstances, if not more so. To enter this air space was asking for a hero's death, and perhaps that is what did happen.

Naturally, the Johnson family were distressed by these slanders and employed a legal friend, Sir William Crocker, to try to establish the facts – in wartime, not an easy task. He saw at once that the only final refutation was to show that there never had been a second body in the water, that the whole rumour was based on the sighting of 'nothing more significant than a small pigskin valise'. To do this he would have to find a sailor who at first had mistaken an item of Amy's luggage for the head of a man, and had then, as it or he came closer, realised that it was only a case.

In these endeavours Sir William succeeded only in annoying the Navy. Lieutenant O'Dea of the *Haslemere* could not be shifted from his statement that he had seen two bodies in the water, a man and a woman, but Sir William claimed that as far as identifying sex was concerned 'he has admitted to us that he was relying entirely upon imagination.' Sir William must have been pressing the officer very hard, for his written testimony shows that, while he was not near enough to the stern to hear that thin, weak voice calling up at the ship from the water, he heard the voices of his own crew shouting: 'It's a woman!' That was not imagination. Some had been close enough to throw lines, one had actually held out his hand for Amy to grasp if she could.

Sir William would have none of it:

> Others who were present confirm, but without any real conviction, that they saw two bodies and having said so once they are all disinclined to permit any modification. The senior service is the senior service and their stories have this in common with the Laws of the Medes and Persians, they alter not . . . For success in an inquiry of this character one needs luck and persistence. I have been persistent enough to earn some unpopularity with HM Navy.

The lawyer's difficulty was that the evidence would not bear his interpretation and all his endeavours to get the witnesses to change their stories failed. With the dead Fletcher he could let his own imagination have free rein, however. Arguing that a man so brave and determined would not give up the rescue without good reason, he suggested that he did so because he found it was a bag and not a person. But, equally, one would abort a rescue if the person was found to be dead, or if one was dying oneself and now incapable of giving help to others, which was in fact Fletcher's situation a minute or two after reaching what was clearly another drifting body, which soon ceased to float.

One of Sir William's partners, Mr Vernon S. Wood, JP, writing to Amy's parents in May 1941, took a line much more in touch with the evidence. He thought the second body 'might easily turn out to be some flotsam from the 'plane itself or even the body of some ship-wrecked sailor or German aviator who had been brought down in the Channel.' If he had just added Italian or British airmen, then his list would have completed the possibilities. Eventually the senior partner had to admit defeat. He wrote to Mr Johnson in July 1942 that: 'none of the evidence I have obtained in any way modifies the misconception of the principal eye-witnesses that there were two bodies in the water.'

But of course there were two, both wearing head-coverings. Amy had a flying helmet on, the man could have been wearing anything – a helmet if he was aircrew, perhaps a woollen balaclava helmet or cap-comforter if he was a seaman or soldier. He could have been long dead, or recently dead, or just dying. No one will ever know who or what he was. That Amy died under the hull of the *Haslemere*, as the trawler's stern fell into the trough of a wave on top of her, is reasonably certain; she would have been either killed outright or knocked unconscious. Her body was never found. Female bones washed up at Herne Bay in 1961 momentarily posed another 'Amy Johnson riddle', but were later shown to be the remains of someone else.

The mystery of why Amy was over the Thames Estuary at 3.30 pm instead of at Oxford 100 miles to the northwest by about 1 o'clock will never be solved. Presumably she miscalculated the winds, became badly lost, and because the aircraft was not fitted with radio could not call for help. Her best course then would be to look for a hole in the clouds until her petrol was running low but was not completely exhausted; an alternative would be to let it run out and then jump by parachute. At 3.30 pm her petrol would indeed have reached the danger point. But then, as we know, she was over the entrance to the Thames Estuary, in the vicinity of a convoy which had just been heavily bombed for an hour. This would be a very dangerous thing to do. Most gunners at sea, and some on land, were ignorant of aircraft recognition. Naval gunners were notoriously trigger-happy and some officers ordered shoot on sight at any aircraft approaching their ship from any direction. They once shot down a Sunderland flying boat at Spithead, just on principle. Amy's plane could have been hit and damaged a few minutes before it was first seen coming out of the clouds, and the gun need not even have been a naval one; the Estuary was well defended by land batteries too.

The fiftieth anniversary celebrations of Amy's Australia flight from Croydon brought a suggestion to a local paper that evidence of this might exist. The writer of the letter, Stuart V. Tucker, stated that a relative of his who had served during the war in the police force had heard that 'examination of parts of Amy Johnson's aircraft revealed damage by anti-aircraft fire.' On inquiring further, I was told by Mr Tucker that the relative had died in 1961 after giving only verbal evidence. However, as he was by no means a 'line shooter', Mr Tucker was inclined to believe that this story was one more wartime 'secret' not yet acknowledged. As the historian of the 'Coal-Scuttle Brigade' (the wartime Channel convoys made up mostly of colliers) I was already well aware of how very likely this possibility really was.

Another suggestion seemed equally plausible. Skipper Williamson of the drifter *Young Jacob* had already pointed out that the *Haslemere* was acting as 'Ack-Ack' ship to the convoy and was towing a balloon. If Amy, almost out of petrol and perhaps fearing she might have strayed out over the sea, saw a balloon above the clouds, she might well assume that it marked the position of solid land and so decide to glide down through the clouds near it. That the balloon might mark the position of the AA ship of a convoy which had just been attacked from the air is about the last thing she would think of.

These suggestions fit the facts well enough, except that it is hard to reconcile them with the statements of several men from the *Haslemere* and one from the *Berkeley* that a person on a parachute came through the clouds first, followed by the plane, which then circled the descending parachute. The plane certainly was the Oxford, no doubt about that. But who was the person on that parachute? The assumption has always been that it was Amy Johnson. It is just possible that the plane might have flown itself in a circle after the pilot had got out. The witnesses varied from half a turn to three complete turns; but the two nearest to the plane were sure it made either two or three turns. That seems odd for an empty aircraft. Could it be that Amy at the last moment caught sight of a parachutist and followed him down through the clouds, circling the spot until she herself ran out of fuel and had to crash-land on the water? The means of escape from the Oxford would still be the same, the pulling of a lever which released a panel large enough to allow a bulkily clothed person, with parachute strapped on, to get through.

That none of the witnesses heard either AA fire or machinegunning at the time the parachutist appeared is no material objection, had the man baled out high up some miles away, drifting slowly down with the strong east wind. I still have a vivid recollection of our ship standing off the coast of Normandy and spending some minutes watching a man on a parachute drifting down very slowly towards the beach defences. We neither saw nor heard the aircraft from which he must have jumped, we were conscious of no gunfire one could tie up with this particular incident, but there he was, whoever he was. And we never found out either.

A last possibility, for which however there is no evidence, is that Amy might have been shot down by an enemy aircraft still ranging the area. One point which is now virtually untenable is the story of 'Mr X'. I contacted Dr Cody, the MO of HMS *Berkeley*, who had gone in the whaler which retrieved Amy's luggage from the water. He told me: 'There was no other baggage of any sort, which, in my view, makes it unlikely that there was a passenger.'

Chapter 7

THE MAN WHO WAS NEVER MISSING
Helmut Wick, J.C. Dundas and ?: August–November 1940

Oberleutnant Rudolf Pflanz pulled the Messerschmitt 109 out of its dive at 3000 metres and headed for home. 'Home' now for the pilots of JG 2, the new Richthofen *Geschwader*, was an airfield in conquered France. The German fighters had become scattered during a high-level dogfight over the Isle of Wight that November afternoon, and Pflanz was alone. The Channel, 10,000 feet below him, was a nervy no-man's-land. Pflanz was well aware of that. He banked quickly, first to the right, then to the left, scanning to make sure that all the air behind him was 'clean'. No Tommies – no Spitfires. Good. There was nothing to fear from Hurricanes, they were too slow and clumsy, but the Spitfires could be deadly. Thankfully, the German pilot set a straight course. He had barely enough fuel to reach the airfield, and any zigzagging would put him into the icy winter sea.

Ahead, 2000 metres further southwards, Pflanz saw two other fighters, also diving for home. That was better still, to have company, for in the vastness of the sky three pairs of eyes were much safer than one. Pflanz advanced his throttle to catch up and had closed to within 1500 metres of the rearmost plane of the pair, when to his horror he saw that it was firing on the one in front! One Messerschmitt attacking another. The number two shooting at the leader? Pflanz rammed the throttle forward, but before he could close the distance the leading plane began to fall, while the fighter behind it fired another burst. This time, Pflanz was near enough to see the watering-can effect of tracer bullets from an eight-gun battery. A British plane!

The Spitfire flew on to the south until the falling Messerschmitt hit the sea then, satisfied of the kill, turned back for the white cliffs of the Isle of Wight. Unaware, the enemy pilot was making interception easy. Pflanz swept in behind him in a climbing turn, caught a momentary glimpse of a German parachute outlined against the sky above, and knew that he could do nothing to help. He could only avenge his comrade. Closing until he could almost see the rivets, he put his first burst straight into the cabin. The British machine went out of control and dived into the sea. Both wings came off at the impact, in a circle of spray, not far from where the empty German fighter had gone in. But of the lone parachute there was no sign now.

Pflanz dared not search. It was getting dark and his fuel was low. But he could call for rescue over his radio. He thought a moment, then sent a carefully-phrased message:

> Forty kilometres south-west of the Isle of Wight, German fighter pilot parachuted next to an English pilot.

That was the truth, although the British airman was almost certainly beyond rescue. But it gave his comrade a double chance. If the German air-sea rescue service could not get there in time, or at all, the English might succeed – the position was nearer to their side of this dirty Channel which, unlike the earth, swallowed pilots without trace. Pflanz, like most of his comrades, hated that deadly stretch of water. There might be no rescue at all. Although it was still light up here in the heights, the chill of winter's dusk already lay on the waves below.

Pflanz kept repeating the call for rescue all the way to the French coast, but it seemed at first as if he was talking to empty air, although friend and foe must be listening. When at last a German air-sea rescue boat called him, it was too late for Pflanz to radio clear directions. His engine stopped. Out of fuel and still far away from the aerodrome! He got down in a field, and set off to walk to the nearest village in search of a telephone. Meeting a French cyclist, he borrowed the man's machine and pedalled off to look for the local Feld Kommandantur. The august individual he at length unearthed, representative of the Third Reich in that village, turned out to hold the rank of corporal. His telephone did not have much priority and it took a long time to get a connection to the airfield, where rescue measures might be prepared. Now it was dark and cold, colder still on the sea, and every minute might be precious. At length, Fighter HQ answered.

They sounded very worried. Half the four-strong staff flight of the Richthofen *Geschwader* had failed to come back, so something serious had happened over there. Leie and Fiby had returned, but Wick and Pflanz were long overdue. Glad as they were to hear that Pflanz was safely down somewhere in France, the news remained black, for the missing pilot could only be Major Wick, the *Geschwader* Kommodore himself, one of the Luftwaffe's three leading aces.

Helmut Wick was born at Mannheim in 1915, the son of an agricultural engineer and a shopkeeper's daughter. For a time the family lived in Southwest Africa. In 1940 Wick was twenty-five, slightly younger than his former instructor Werner Mölders, aged twenty-seven, and the other leading German ace, Adolf Galland, twenty-eight – a similar age group to the fighter leaders of 1914–1918, for in the year of their deaths Boelcke was twenty-five, Immelmann and Manfred von Richthofen both twenty-six. The age difference meant that while Mölders and Galland had gained invaluable combat experience in the Spanish Civil War, Wick was still in the training stage. Yet by the end of 1940 the three top scorers were Galland with fifty-seven victories, Wick with fifty-six and Mölders with fifty-five. Admittedly, only one of them was still alive.

In January 1939 Wick was a young officer in the Staffel led by Mölders. He saw his first real air fighting during the winter of 1939–1940, but his meteoric rise began with the German armoured onrush into France on 10 May 1940. In eleven airfights he scored thirteen victories. His foes included aircraft of the French air force – Curtiss Hawk, Bloch and Morane-Saulnier fighters, Potez 63 fighter-reconnaissance machines and Leo 45 bombers. His bag of RAF planes included oddities such as a Westland Lysander, a Vickers Wellesley and a Westland Wapiti biplane, as well as a Hereford, a Blenheim and three suicidal Fairey Battles over the Marne Bridges. Then the Battle of France was over and Wick found himself on the Channel coast opposite Portsmouth, Southampton and the Isle of Wight.

The front over which the new Richthofen Circus was to fight the RAF contained not the evocative French and Belgian place-names of the Great War and its fixed trench-system, but homely English words standing for aerodromes and aircraft factories, including the Supermarine works at Southampton, building and repairing Spitfires. The Luftwaffe was tactically divided to support the army and it was Hugo Sperrle's *Luftflotte* (Air Fleet) 3 which now had the task of putting on a series of air displays over the area which were to make the Schneider Trophy contests appear very minor spectacles. In addition to more than 330 twin-engined bombers, Heinkel 111s and Junkers 88s, Sperrle had in Air Fleet 3 some 250 Stukas (Junkers 87s) of VIII Flieger Korps, commanded by Lieutenant-General Wolfram Baron von Richthofen, a close relative of Manfred, the Red Baron, who had served with the original 'Circus' in France in 1918. Sperrle also had among his three *Jagdgeschwader* the reconstituted Richthofen Circus, JG 2. In the course of the next few months Helmut Wick was to rise from junior leader to command of the whole *Geschwader* in the battles over the Channel coast.

In July he had got a Blenheim off the Isle of Wight and later a Spitfire, but to run up a big score you needed large-scale and continuous action and this did not begin until August. Even so, Wick's first experience of a 'Big Strike', the '*Grosseinsatz gegen England*' ordered for 18 August, proved disappointing. That day Sperrle committed more than a hundred Junkers 87 dive-bombers of Richthofen's Stuka Korps on a thirty-mile front from Portsmouth to Poling in Sussex, with more than 150 fighters as escort. Wick flew with the fifty-five Messerschmitt 109s employed in a 'free hunt' to sweep the skies of RAF fighters above and beyond the target, with complete freedom of aggression, while other 109s had the thankless task of flying close escort on the slow bombers below them. But Wick was bored, for everything went too well. He wrote: 'We didn't find anything interesting over England, so as a kind of compensation we threw ourselves at the "inflated competition" – six balloons.'

I can confirm these victories. Not only did I witness them at close range (one balloon falling 200 yards to my front), but I recorded them with camera and notebook, without however knowing at the time who these particular Germans were, let alone suspecting a connection with Richthofen. Although I had hoped to become a fighter pilot myself, at the outbreak of war the aero clubs had been disbanded and their training aeroplanes towed away to be stored in garages. Neither the aeroplanes nor the pilots were needed now there was a war on, for the Air Ministry had the matter magnificently in hand. Eleven months later, however, the Luftwaffe was blasting RAF property to pieces in front of me, entirely without opposition from the RAF. I had just seen Gosport aerodrome destroyed in three minutes by Stukas, the fifth dive-bombing attack I had witnessed and the fourth in a week. Then Wick and his companions swept in to cover their withdrawal. This is what I wrote within minutes of the event:

> A lull; then the sounds of invisible aeroplanes, and canon and m.gs. Three or four fighters were diving at the Portchester balloons. One balloon blazed, fell slowly, gleaming red and orange, with a black plume above it. Like sharks the flight of black Messerschmitts rolled and plunged amongst the balloons. Over the dockyard another gasbag flamed immediately, and trickled down the sky like a fiery blob of water running down a window pane in slow motion. Above the woods near Cams Manor, ¼-mile down the road, a third balloon trailed dismally earthwards in flame and smoke. The pop-pop-pop of canons and the rattle of m.gs. sounded. A fourth balloon blossomed into flame. The Cams Manor balloon landed two fields away like a blackened and blazing parachute. I took photos like a machine-gun, so when Colin (my brother) told me that a small boy had recovered a canon shell two yards from where I had been standing, I was most surprised. I could count six smoke columns in the sky. I saw four balloons go down in flames personally. No Me. 109s fell. No Junkers came down, that I could see. I'm dashing in and out of the house to write this up, have just come out in time to see three black streaks suspended in the sky over Gosport, with a little glow of fire at the end and moving slowly downwards. With the usual noisy death music. Nine balloons down for good. No German losses over here. The burning aerodrome still pouring up clouds of smoke, grey-white in colour.

This beat Heinrich Gontermann hollow, and Frank Luke and all the other First World War balloon-busters as well. I'd read all that years before in the 1930s, and the writers always described how the balloons 'exploded' in the face of the diving pilot. They quite definitely didn't, so much so that I added a postscript: 'Funny how balloons wriggle down like worms, ever so slowly, twisting and writhing from side to side, lurching in slow motion to their doom.' Very little remained of the balloon except the charred mooring ropes when it fell to the ground, but the single wire mooring cable was left draped over chimneypots, trees and fields.

A genuine difference between then and now was that the old aeroplanes flew so low and slow that ground watchers could make out the red of a Richthofen Circus Albatros or Fokker. This was not so now. The 109s of the new Richthofen *Geschwader* were often so high in the hazy south that they were invisible and only the sound of their guns told you they were up there; then when you did see them it was in silhouette, so that they appeared black. Whereas the line of the trenches in the Great War had given an east-west confrontation, this time the Germans came at you from the south, out of the sun, and at three times the speed.

Of course, the fighting was spread over a great area and you could not see all that happened. On this day the Richthofen boys were still hanging about over the Channel, having just downed the balloons, and with fuel to spare, when Wick and Pflanz saw an Me 109 limping from England with a damaged radiator trailing vapour. 'Hier stinkts eine Me, let's stay with him,' called Wick. They escorted their comrade back until, still fifty kilometres from France, the damaged machine was down almost to wavetop height. It touched, bounced back into the air, then turned tail over prop, throwing the pilot clear. The spreading green stain on the water, from an opened marker-dye bag, was proof that the man was conscious. That was almost a miracle in itself, but the second miracle occurred almost at once, when Wick spotted a Heinkel 59 air-sea rescue seaplane patrolling the flight path of the retiring German formations. He managed to attract its attention and the ambulance plane was able to put down on the sea alongside the downed pilot and fish him out of the water. There was to be nothing as efficient as this until long after the war when the RAF and the Navy began to use helicopters for air-sea rescue.

It was against RAF Fighter Command, not against inferior air forces, that Wick rapidly ran up his score. The Luftwaffe had an accelerated promotion system to 'push' successful men: Wick was given the Knight's Cross at the end of August, promoted Hauptmann to lead I Gruppe of JG 2 on 4 September, fought in the Battle of London from 7 September on, and in the first week of October was promoted Major and given the command of the entire *Geschwader* of more than a hundred fighters. Up to now, the post of Kommodore had been held by a full colonel or half-colonel, a much older man. By the end of October, he was still scoring in spite of the extra load of paperwork to be done at night after flying; on the 29th he shot down numbers forty-three and forty-four. Then a lull due to bad weather, two more victories, followed by the events of 6 November when the *Geschwader* took part in 'a real big circus' as Wick's biographer described it. Wick himself wrote:

> Today we had a terrific time again. We meet a heap of Hurricanes which fly lower than we do. I am just getting ready to attack when I notice something above me and I immediately shout over the radio: 'Attention! Spitfires above us!' But they were so far away that I could start to attack the Hurricanes. They were just turning from their original course and that sealed their fate. Almost simultaneously, the four of us (the staff flight) fired at their formation. I accounted for one. The rest of the Hurricanes moved away, but then pulled up higher. During this manoeuvre I once again caught the one on the right-hand side outside, and he was done for immediately.

This dogfight or a similar one exploded over our house at the back of Portsmouth harbour, with the rival fighter squadrons making vapour trails against a bright blue sky. I took two photographs as the German attack developed, and one after it was over, with the trails fluffed out. The first two pictures show four fighters in two pairs turning close together, chasing other machines in unison, while high above them is a vic of three machines, also turning. German fighters flew in gaggles of loose pairs, unlike the British who still used the pre-war V-formation, well-suited to air displays but not to war. I also logged the affair and can see that date, time and description fit Wick's narrative; the only discrepancy is that JG 2's kill book shows the place as Southampton instead of Portchester, which is northwest of Portsmouth towards Southampton. So, unlike 18 August, I cannot be one hundred per cent sure that I was photographing Helmut Wick in action. My diary read:

> *Afternoon.* Sirens, then gunfire. I spotted many vapour trails over Portchester. There seemed to be several formations. First came 4 planes travelling westward towards us under the trails. 2 silver specks danced above, and were joined by 3 more. Trails formed on some of these. 8 planes in compact formation left comet-like trails. Gradually a confused mass of white vapour

trails, like an octopus all tangled up, came into being above our heads as the fighters weaved in and out. A plane stood on its ear; another looped, like a coin spun casually upwards from the fingers; others whirled round each other in merry circles to the chattering tune of m.gs. Then the sky rained planes. Like the leaves in autumn they came to earth. A Hurricane shot over Wallington very fast and at a flat angle, to disappear behind the hill. Almost on its heels came another plane, falling fast and flatly. Neither rose again. Colin yelled: 'Look – he's going straight down. He's into a cloud now. He's doing a terrible speed . . .' I looked, and though I failed to see the vertical gent, I spotted a plane descending just above the horizon. A new sound cut in above the roars of diving fighters and the fierce, deadly chatter of machine-guns. A Hurricane came over from behind the hill low down and banked overhead. He stood out against the confused tangle of trails, tearing across its pattern at low height. It was strange to think that men were dying up there . . . that the little silver shapes contained men who were trying – not without success – to kill each other. Eventually, some comets reformed and headed south.

Wick records how, after scoring two kills off the Hurricane squadron, picking off the outside man each time, he got cold feet and headed for home a few minutes before he really had to; and that on the way south, he jumped a squadron of Spitfires and shot down three. That was too far away for me to see, but his description of the second Hurricane he hit, that 'he was done for immediately', fitted the 'vertical gent' I failed to see. I failed to see him either because of cloud or because he was not where the noise was. It was a most terrible screaming, howling roar – he was going down vertically with engine full on probably doing at least 500 mph, and no doubt the pilot was dead already.

My brother and I got out our bicycles and set off uphill to find the wrecks, which is not as easy as one might think. We asked our way from witness to witness. Some talked of a 'yellow-nosed Messerschmitt' that had been shot down directly overhead (probably the machine that had gone down on the heels of the first Hurricane). A woman said she had bussed past a burning plane near Widley. An old man had seen a black plane skim a wood at a flat angle, and two other men hadn't seen anything at all. We reached the crest of Portsdown Hill – and there it was.

Andrée, Fraenkel and Strindberg in the car of the Polar balloon just before lift-off on 11 July, 1897. *Photo: Andréemuseet, Gränna*.

The balloon *Örnen* leaving Dane Island with sails set and towing draglines. *Photo; Andréemuseet, Gränna*.

Above: The *Eagle* has landed and will fly no more. If the aeronauts are to escape, they must walk to safety across the pack ice. *Photo: Andréemuseet, Gränna.*

Centre: Fraenkel and Strindberg have shot a bear during their march across the ice. Each one, they wrote, was "a walking butcher's shop". *Photo: Andréemuseet, Gränna.*

Below: A B.E.2 similar to the machine in which Desmond Arthur was killed flying over the Royal Aircraft Factory at Farnborough before the 1914–18 War. *Photo: Author's collection.*

Baron Manfred von Richthofen, highest scoring ace of the Great War of 1914–18. *Photo: Author's collection.*

The Richthofen Circus in France, 1917. *Photo: Author's collection.*

The Richthofen Geschwader in France, 1940. *Photo: Author's collection.*

German Observation balloon on the Western Front, 1914–18. *Photo: Author's collection.*

Lt. Luke with a German plane which he shot down near Verdun on 18 September, 1918. *Photo: US Signal Corps 111-SC-24397, National Archives, Washington.*

Lt. Frank Luke, the American balloon-buster. *Photo: US Signal Corps, National Archives, Washington*.

F/Lt. S. M. Kinkead. The original painting hangs in the Officers Mess at the R.A.F. College, Cranwell. *Photo: R. J. Mitchell Hall, Southampton*.

F/Lt. D'Arcy Grieg taxis a Supermarine S.5 past one of the Italian contenders, a Macchi, during the 1929 contest off Southsea. Note how the spray blown up by the propellor envelopes the cockpit. *Photo: A/Cmdre D'Arcy Grieg's collection*.

The fate of many an ocean flyer. Commander Byrd survived this crash in the English channel in 1927. *Photo: Author's collection.*

Amy Johnson.

The author at Portsmouth Aero Club in 1933, aged 15, after dual instruction in a D. H. Cirrus II Moth, the direct descendant of the B.E.2. *Photo: Author's collection.*

Sir Charles Kingsford-Smith (on right) and Captain P. G. Taylor with their Lockheed Altair *Lady Southern Cross*. *Photo: Author's collection.*

Major Helmut Wick, leader of JG-2. *Photo: Author's collection.*

Wick starts to burn the Portsmouth balloons on 18 August, 1940. Wartime watercolour sketch: *Author*

Above: Two more balloons go down before the Richthofen Geschwader on 18 August, while Gosport aerodrome burns fiercely in the background after stuka attack. The nearest balloon, almost burnt out, is 200 yards in front of the camera lens. *Photo: Author*.

Below: Wick and his staff flight attacking a Hurricane squadron over the back of Portsmouth harbour on 6 November, 1940. *Photo: Author*.

Right below: Author's wartime watercolour of the Hurricane crash-site at Widley, 6 November, 1940. Note sentries and ambulance as well as Hawker Hart.

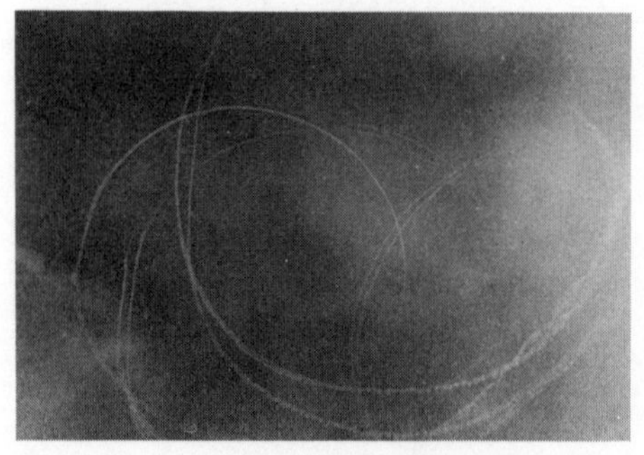

Above: The Hurricane crash-site at Widley being dug on 6 October, 1979. By coincidence, the Wealden Group's vehicle is identical to the military ambulance which entered the same field on 6 November, 1940. *Photo: Author.*

Centre: Looking downhill across the excavation to Mr. Ware's farm. *Photo: Author.*

Left: Ernest Ware, the farmer who saw the Hurricane dive into his field in 1940, with one of the recovered Browning guns in 1979. *Photo: Author.*

Left: The excavation at Widley goes deeper. The pilot's remains are put into a black polythene bag to be kept separately. *Photo: Author.*

Below: Scorebook of the Richthofen Geschwader for 6 November, 1940, showing claims by Wick, Leie and Pflanz.

Author's wartime watercolour of the contrails high up over the Isle of Wight on 28 November, 1940, as the Spitfires of 152 and 609 Squadrons engaged the Richthofen Geschwader a few minutes before Major Wick was shot down.

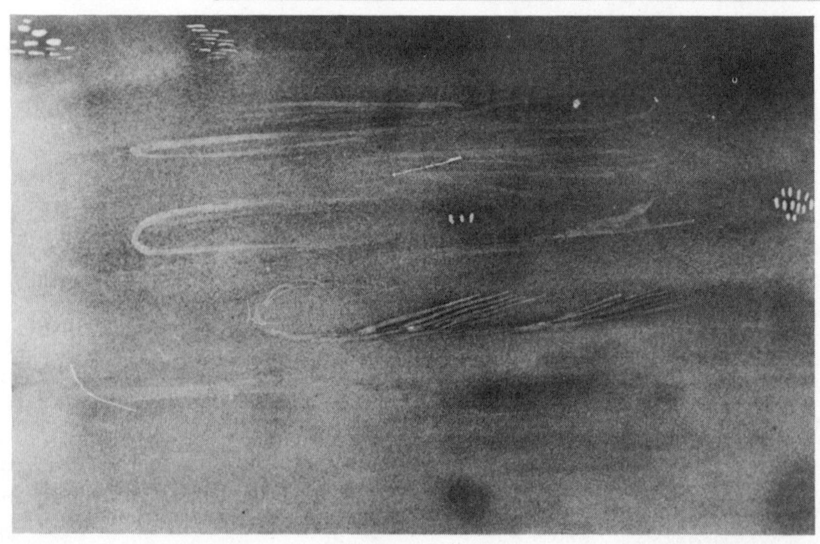

Above: 'Hugh', Violet and Garry Brown outside the window where 'The Pilot' is seen. *Photo: Author.*

Below: Author's wartime watercolour of the Me 110 which was brought down by a Hurricane at Ashey. Both crew members landed alive, but the pilot was shot by the Jersey Militia.

Below: Heinkel 111h of KG-26, the Lowengeschwader, carrying a 2½-ton bomb externally. *Photo: Willibald Klein's collection.*

Above: 'A big, black gaping smoking hole, surrounded by litter and complete death and devastation' – the crater in Conway Street, Christmas, 1940. *Photo: "The News", Portsmouth.*

Right: A prewar portrait of Mrs. Lilian ('Dolly') Owen, who was stripped naked by the blast from the Conway Street bomb. *Photo: Mr. & Mrs. Owen.*

Below: Nineteen little streets were destroyed or made uninhabitable by one bomb. *Photo: "The News", Portsmouth.*

'When daylight started coming, then you could see that the ruins were smoking, were very hot' – and there were people trapped underneath. *Photo: "The News", Portsmouth.*

The Conway Street area on 21 December, 1979. The only building which remains from 23 December, 1940 is St. Agatha's Church. Everything else has been levelled to make a dockyard car park. *Photo: Author.*

Above: Joe Kennedy Junior (standing, extreme left) and the crew of his Liberator of Squadron VB-110 at Dunkeswell in March, 1944. *Photo: Cdr. John F. Burger.*

The crew of HMS *Centurion* being taken on board the control destroyer HMS *Shikari* before target practice. *Photo: W. J. Wallis collection.*

The battleship HMS *Centurion*, now crewless, being steered by radio control as a target ship in the 1920s. *Photo: W. J. Wallis collection.*

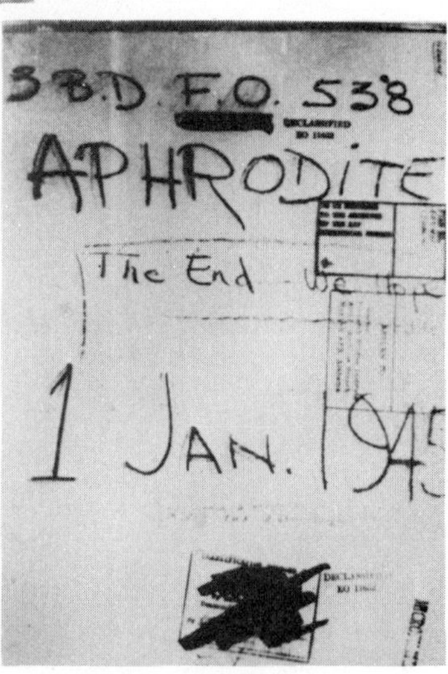
33.D F.O. 538

APHRODITE

The End

1 JAN. 194

Top: Author's notebook sketch, done in September, 1944, of a much-bombed V.1 launch site near Dieppe which had just been captured by the First Canadian Army.

Centre: Author's notebook sketch of September, 1944, showing that even a very near-miss on the control bunker of a V.1 site appears to have been ineffective. The site stopped firing because the Canadians captured it.

Right: On the last page of the Aphrodite project file some embittered American has written 'The End – We Hope'. *Photo: Stewart Evans.*

Stewart Evans with a battered engine from Kennedy's disintegrated Liberator. *Photo: Dennis Eisenberg.*

Oblt. Karl-Heinz Willius, CO of II Gruppe, JG.26. *Photo: RNethAF Collection*.

A Hart was circling Widley – like a vulture floating round a corpse. In a field below the road, a group of tin-hatted men stood around a white gash in the soil. I biked through a field and turned the glasses onto a small, elongated tangle of silver and green metal. In most places it wasn't more than six inches high. Smoke drifted faintly away on the wind. One chap had seen it come down. It simply dived straight in under full throttle, he said. It had been a Hurricane. The ambulance was waiting for the soldiers to pull the pilot out – but the smouldering heap was too hot to touch. I should say he was ground to powder anyway. Small boys had collected bits of metal, and their pockets bulged . . . We rode off down the hill, met a boy with a delivery van who had seen a British plane down in a field near Denmead, and heard a man talking about a 'yellow-nosed Messerschmitt'. Near Denmead, a girl gave us directions to get to a crashed plane near the Forest of Bere. Everyone we met, when told there was a plane down at Widley, said: 'Oh, good – another one the less.' They cannot conceive that ours can be shot down, too . . . I suppose we rode 40–50 miles to find the wreck of a Hurricane and prove the existence of another, plus a 'yellow-nosed Messerschmitt' and an unidentified 'possible'.

In the privacy of my home I did a watercolour sketch of the crash scene at Widley (sketching, taking notes and photographing the war being publicly forbidden), so the evidence remains. There are the army sentries with bayoneted rifles in a ring round it, the waiting ambulance, the circling yellow-painted Hawker Hart above, men in RAF blue and one man in black, perhaps the village bobby, plus miscellaneous spectators lining a hedge. Thirty-nine years later, that sketch was to help identify the crash site. Incredible as it may seem, the pilot of that Hurricane was listed as 'missing', and still is.

The target of the raid which JG 2 were escorting that day had been Southampton, so that may explain why Wick entered 'Southampton' in the score book. Next day, 7 November, he claimed a Hurricane at 7000 metres over 'Portsmouth' but that action was too far south of Portsmouth for me to see more than two under-strength squadrons which by their formations appeared to be British. RAF records were to show that Hurricanes showered down, mostly well south of Portsmouth in the Isle of Wight area. No. 145 Squadron lost no less than five machines, one near Ashey

Down in Wight, two just offshore; while a fourth fell at Wittering in Sussex and a fifth was unaccounted for, vanished without trace. A Hurricane from 56 Squadron went into the sea just off Woody Point in Wight. All victims in the same battle to the 'yellow noses' of the Richthofen *Geschwader*.

On 28 November there was continual action by Wick and his men, both in the morning and in the afternoon, which was visible from the back of Portsmouth harbour. I had the benefit of excellent German naval fieldglasses taken out of the destroyer *Hans Ludemann* at Narvik early in 1940 and lent to me for the Battle of Britain, of which this was virtually the concluding day.

In the morning JG 2 were covering fighter-bombers ordered to attack one of the Southampton aircraft factories. These flew faster and higher than purpose-built bombers, so when Wick spotted a squadron of Spitfires making trails at an extraordinarily high altitude, he was alarmed. They were well-placed to knock down the attackers while the Germans were still hampered by their bomb-load. Putting on boost, he outclimbed the three '*Katschmareks*' (faithful followers) of his staff flight in his haste to reach the British squadron. At 11,000 metres, well over 30,000 feet, Wick found himself alone with all those Spitfires, but attacked unhesitatingly. Pflanz saw one Spitfire burn and begin to fall, saw the pilot come out and a parachute open. In the selective lens of my borrowed fieldglasses I saw one vapour trail and a tiny silver speck losing height above it and then crossing below, probably Wick and his victim.

Sandy Johnstone of 602 Spitfire Squadron was also keeping a diary at this time:

> Mickey took the squadron on patrol to tackle a raid approaching the Isle of Wight very high up. I don't know whether the Ops people were merely trying to impress our distinguished visitor (the Commander-in-Chief, Sholto Douglas), or whether they had had a sudden rush of blood to the head, but the lads were ordered to 33,000 feet where they were well and truly out-manoeuvred by the 109Es. Pat Lyall was shot up but managed to keep up with the others on the return journey for most of the way when, for some inexplicable reason he baled out at low level, when his parachute failed to open. What rotten luck.

In spite of the discrepancy between the parachute opening or not, 602 must have been the high-flying British squadron. It was Major Wick's fifty-fifth victory. On the Normandy air base, however, there was a lively debate about the new British tactics. There was no suggestion that the RAF had had 'a rush of blood to the head'; instead there was alarm. 'Until now they have never met British planes at 11,000 metres,' wrote the journalist commissioned to write up Wick's exploits. 'Only later was it discovered that this British unit was led by von Mader who was considered to be the best English fighter pilot.'

At 602's Westhampnett base, they were also debating the matter. 'There was a blazing row about Thursday's action,' wrote Johnstone who like the others had previously agreed not to go so high because they thought the latest mark of 109 superior to the Spitfire at that height.

The 'von Mader' the German journalist referred to was probably Douglas Bader, who was one of the best-known British pilots at that time and after. 'Bader' is originally a German name, of course. Oddly, one of the Tangmere squadrons at this time, No. 607, was led by Jimmy Vick.

The weather remained good but there were no bombers to protect, so Helmut Wick ordered a sweep which, based on the experience of the morning, must be carried out at the highest altitude possible. I saw that air battle in the far distance as unfolding patterns of vapour trails in the southwest which, totted up, meant at least sixty fighters engaged; and I made a few sketches of that sky-writing. I did not know that a new Richthofen Circus was up there. I had never heard of Helmut Wick, and I did not see his end because that was far to the south of the Isle of Wight, and the only witness was Oberleutnant Rudolf Pflanz, whose narrative begins this chapter, but which I did not know about until nearly forty years later. However, German broadcasts and newspapers were being monitored in England and the story filtered through to the British press.

The first newsman to realise that a new legend had been born and a new mystery created was Noel Monks, air correspondent of the London *Daily Mail.* Headlined 'THE PILOT NO ONE KNEW : KILLED NAZI ACE NO. 1', his report read:

> The name of the R.A.F. pilot who shot down Major Helmuth Wieck, (sic) Germany's greatest air ace, will never be known. The pilot died in a grim battle off the Isle of Wight on November 28 against a flight of M.E. 110's led by Wieck. The Hurricane pilots, outnumbered, fought until one by one they were shot down. One of them shot down the Nazi ace and was then himself shot down by Wieck's best friend. Major Wieck was the leader of the Richthofen Squadron. The Nazis claim he had 56 British, Polish, and French aircraft to his credit. An R.A.F. officer told me that Wieck's reputed 'bag' of 56 was probably correct. He said: 'We had heard a lot about him, but none of our pilots knew when they engaged him in battle. Unlike the Richthofen Circus in the last war, when Richthofen flew a red plane, Wieck's squadron all flew aircraft with the same distinguishing marks – yellow noses.'

On 26 April 1941, the Portsmouth *Evening News* carried a release from the Air Ministry News Service, which began:

> The name of the British pilot who shot down Germany's greatest fighter ace, Major Helmoth Wieck, Commander of the notorious Richthofen Squadron, was disclosed today. He was 25-years-old Flight-Lieut. John Charles Dundas, 609 (West Riding) Squadron. Wieck was his thirteenth known Nazi victim but he never knew the distinction that was his when his guns blazed on the 28th of last November and sent one of three Messerschmitts that swooped on him in close formation spinning down over the Isle of Wight . . . He was in turn shot down by the other two Messerschmitts and killed. One of the section leaders of 609 Squadron reported having seen a parachute with a tear in it going down about 10 miles west of the Needles. Search of the sea lasted for over an hour until petrol began to fail and the squadron was forced to return . . . This is the story of how Goering's crack pilot met his match in a combat of seconds . . . Wieck was regarded as one of Germany's greatest fighter aces. He had fought with the Luftwaffe in Spain and Poland . . . Pilots flying with him on the day he was killed have been interrogated again and again by the chiefs of the Luftwaffe. Broadcast accounts of his last fight have been made from many German stations but the true manner of his death has never been fully revealed.

Dedicated error-spotters and propaganda-buffs could have a field-day with those two pieces, but it is true that there was a mystery: both sides hoped that their man had been picked up and taken prisoner, rather than swallowed by the cold Channel. But as the days and weeks went by, hope faded. On the morning of 29 November, the Germans launched an intensive search to within sight of the Isle of Wight. Oberleutnant Pflanz led the air search, naturally, and the German navy supported him with E-boats and other craft. It was a misty morning. Pflanz tried to duplicate the height and distance flown from England the previous afternoon, 'but in the area where the actual tragedy took place, there was nothing to be seen.' Allowing for the actions of the tidal streams, Pflanz soon picked up traces of a large oil-slick. Wick or the enemy fighter? Soon after, he found a second oil-slick, to the south of the first one. The Spitfire had gone in first, so this southerly stain on the sea must be from Wick's Messerschmitt. They searched around it for two hours that morning, assisted by the navy, and again in the afternoon, but of the lost pilots there was no trace. They had vanished.

Two biographies of Wick appeared in 1941: a booklet published by Walter Zuerl of Munich, and a book largely written by an official war reporter, Josef Grabler, who was killed in the airborne invasion of Crete before he could finish the work. They too were subject to wartime censorship and propaganda, and their authors had no access to RAF records, naturally. These are now in the Public Record Office at Kew, and show that a number of squadrons were involved, including 609 and 152, both flying Spitfires not Hurricanes, against 109s not 110s.

609 was an Auxiliary Air Force squadron. John Charles Dundas had joined it in 1938 when he was one of the more brilliant of the younger journalists on the *Yorkshire Post*, specialising in European international affairs. His public school was Stowe, his university Oxford, where he had taken a First in Modern Greats. Between 31 May and 28 November 1940, he had run up a score of 13½ enemy aircraft destroyed, seven damaged, four probable, mostly over Hampshire and the Isle of Wight. But on 27 November he had chased a Junkers 88 all the way to Cherbourg on his own, setting one engine on fire and almost exhausting his ammunition in the near-vicinity of a German fighter base. Dundas was now a Flight-Lieutenant with the DFC.

On 28 November, 609 scrambled several times from Middle Wallop, paired with another Spitfire Squadron, 152. Middle Wallop was in the middle of an official Army visit that day, with staff officers and 150 cadets from the Royal Military college being given a guided tour of the aerodrome. 609 itself was packing up, ready to move to another aerodrome next day. From one of these scrambles Dundas and a new man, Pilot-Officer Baillon, failed to return. The only clue had been a single sentence from Dundas, heard by the Controller and by the squadron commander, Michael Robinson:

'I've finished an Me 109 – Whooppee!'

Robinson replied: 'Good show, John.'

After that, nothing more. A German announcement that Major Wick had been killed that day off the Isle of Wight, and that his friend had at once disposed of the British victor, made it seem likely that either the experienced Dundas or the experienced Sergeant Klein, a Pole serving in 152 squadron who also did not return, might have been the man who should be credited with Wick. So they thought in 609, but 152's records tend to eliminate Klein, although they introduce another pilot who scored in the pursuit phase, when the 109s were diving for home, low on fuel, from above the Needles. This was Pilot-Officer Marrs, who had seen a Messerschmitt shoot Pilot Officer Watson's Spitfire, which went down trailing smoke and with Watson trying to get out of the cockpit. His plane and his body were found southeast of Wareham, the parachute canopy rent where it had probably caught on part of the Spitfire. Marrs followed the 109 as it headed for France and gradually overtook it. He got in some good bursts, the pilot baled out and the 109 blew up. So that was one more German down in the Channel. But Marrs got back, so the man he shot down could not have been Wick.

Nor was Klein likely to be the victor, although it seems he too went down into the Channel. Flying Officer Holmes caught a momentary glimpse of a Spitfire upon which he saw what appeared to be 'a bright spark'. This plane then spun three or four times in his view before Holmes had to give his full attention to several 109s trying to kill him. Klein did not return and this burning Spitfire last seen spinning down over the sea was almost certainly his and not the machine flown by Dundas.

Wick's staff flight, after seeing the Major score a final victory over the Bournemouth area – the British machine lost its wings and the fuselage went down like a torpedo, the wings whirling round above – lost sight of their leader. Then a big British formation came down on them vertically, streaming contrails behind them, wing-guns rippling fire at the Germans. Oberleutnant Leie and Leutnant Fiby dived out of it together, but lost sight of Pflanz. Pflanz later sighted a Spitfire which shot down Wick, which he in turn shot down. Almost certainly, this was Dundas, although for a time the optimists in 609 Squadron hoped that he might have repeated his exploit with the Ju 88 over Cherbourg, and been shot down safely over France. Similarly, the Germans asked the British for news of Wick, hoping that he had been safely picked up. But there was no news of the two aces, nor of Sergeant Klein and the unknown Messerschmitt pilot whose plane exploded over the sea. The Channel took them all without trace.

A few came back, however. There were some from 602 Squadron, as Sandy Johnstone noted in his diary. The burned body of one pilot was found in an orchard, with no sign of his plane. Then there was Sergeant Sprague, lost on 11 September:

> We just cannot budge Mrs Sprague. The poor soul sits in her car for hours just gazing towards the B Flight huts as if willing her husband to walk through the door. It is really tragic for all of us, for there is still no news of her husband, and we fear he must have come down in the sea. Nor is there any news of Harry Moody, whose brother has also been in touch with me. Alas, I can offer neither much comfort and I find the whole thing deeply distressing.

Sergeant Sprague came back on 11 October, washed up by the waves on Brighton beach. But so many simply and sometimes inexplicably, disappeared.

The most unbelievable story concerned the Hurricane which dived vertically into Portsdown Hill near Fort Widley on 6 November 1940, the crash site being crowded with witnesses. No one doubted that the pilot was still in it, because no one had seen any parachutes. I included a description of this dogfight, and also the death of Wick, in my book on the Battle of Britain (*Strike from the Sky*, Souvenir Press, 1960). Many years later, when an interest arose in aircraft archaeology, I was visited by Brian Connolly of the Wealden Aviation Archaeological Group and we discussed three possible targets, a Ju 88 dive-bomber I had seen shot down onto the mudflats near Portchester Castle, the Widley Hurricane, and a German bomber I had seen burn up and fall apart at night near Chichester. Of course, it was not so much the aircraft themselves as the story associated with each incident which was being researched; the planes, when found, were just parts of the historical evidence which it was important to secure while eye-witnesses and participants were still alive.

In 1979 Andy Saunders, the secretary of the Group, got in touch and loaned me a copy of the 1941 biography of Wick by Josef Grabler, which he had found on a second-hand bookstall during a visit to Germany. The publishers had chosen to include just one page from JG 2's combat files, and this turned out to be the entries for 6 and 7 November 1940! As Andy pointed out, it looked very much as if Wick himself might have shot down that Widley Hurricane; or, if not Wick, then another pilot from his Group, Leutnant Schell. In any case, they intended to dig the Widley Hurricane. They were fairly certain who the pilot was, but they had discovered that he was still listed as 'missing', and that the RAF demanded visible proof of identity, such as personal discs or an aircraft number of some kind. If they could find these, then the mystery might be cleared up publicly.

The excavation began at the farm of Mr Ernest Ware on 6 October 1979, of what should be the crash site of a Hurricane shot down on 6 November 1940. The connection needed to be proved. My sketch was not enough in itself, although it did match up on the ground with one major discrepancy only – there should have been a hillock to hide the wheels of the waiting ambulance. Mr Ware was able to say that there had been such rising ground but he had bulldozed it level.

Oddly, the Wealden Group had brought with them a vehicle which had been an ambulance and was identical to the one I had drawn waiting by the crash in 1940. At first, I had trouble orienting myself – thirty-nine years had passed since I had last stood in that field – but increasingly I began to feel that it was the right one. To prove that it was, there had to be a Hurricane in it. And not just any old Hurricane, but a Hurricane showing signs of having dived in vertically from 15,000 feet at over 500 mph with the pilot still in the cockpit. In all their considerable experience of excavating crashed aircraft of the period, the Wealden group had never yet found a site where the plane had been really vertical at the moment of impact. But if the engine was at full power at the moment of impact, the sudden stop of all that fast-moving machinery created an explosion which tore the machine to pieces. And this was just what Ernest Ware and his brother Bob were telling us.

They had been standing in a field about 200 yards away with a tractor and cart. The noise of the falling plane was such that they both got under the cart, but Ernest had been looking out in the right direction and saw the Hurricane hit the field with a bang, going vertically. Bits flew up like a fan, then fell back onto the crash site and across the road also. I heard at second hand that one of the pieces was a flying boot, which was thrown back into the hole made in the ground by the Hurricane. Also, there was such a lot of stuff showering down from the dogfight above that Bob Ware seriously doubted my statement that I had taken photographs of it.

One way or another, the Wealden Group's metal detector was not going to be a lot of use in finding the exact spot to dig in that large, sloping field. Their first choice proved to be wrong, by a narrow margin. After the mechanical excavator had removed a layer of recent topsoil, the harder chalk underneath was seen to be littered with small pieces of dural, scattered. They instantly read this as bits of the exploded Hurricane which had fallen back onto the field and subsequently been shifted around by ploughing. On that basis they moved the mechanical excavator to another place nearby, seeing signs not obvious to me.

I was not without experience myself, but my excavating scene was fifty feet underwater at Spithead on the wreck of the Tudor battleship *Mary Rose*, which I had discovered in 1966 after a deliberate search campaign because of the historical importance of that ship, her armament and her crew. Because ship and contents were all unknown, the excavation had to be delicate and slow, carried out over a period of years. A Hurricane was a different matter; furthermore there were three witnesses of the action standing around the 'dig' – the Ware brothers and myself – whereas the last witness to the sinking of the *Mary Rose* must have died more than four centuries ago. I was of course used to finding and photographing the remains of Henry VIII's men and knew that the skulls and bones of the long-dead are not either eerie or horrible.

But for the first ten minutes or so of this dig, it was strange: wondering if my 1940 sketch did really match this field – was it the right one after all? – and the increasing certainty that it was. Then the nagging suspicion that it might be so easy for the Wealden Group to miss the exact spot, because so much light scattered material must litter the chalk, and the rapid reassurance that they really knew their business, with the smooth move to site two, and the discovery of much larger fragments there. Within minutes, a complete group of four Browning .303 machineguns was found – one battery of a Hurricane (the grouping in a Spitfire was different). As the guns were a known distance outboard of the fuselage, the dig was enlarged to the northeast. This uncovered a tangled mass which was part of the tail, together with two human bones about three feet below the field's surface, and then part of the rudder. One foot deeper in the soil the excavators unearthed bits of broken armour plate from the pilot's seat, together with a third bone in association and at this level. The Group's recognition of bones (everything was smeared with chalk) was very fast, and one hardly noticed them being discreetly put into a black polythene bag.

I did not feel morbid or sad. I knew that the pilot was twenty-three when he went into this field, and I was twenty-two when I first came there. When the tail was revealed I could reconstruct the Hurricane in my own mind and people a world with young men, myself among them. The scene as it really had been. I shed forty years in a few minutes and for a brief space was really in contact with the past, so that I was twenty-two again. What happens when we tell war stories over and over, ourselves, or see them told on the 'box', or read them in books, is that no matter how skilful the story teller, he is putting down one more layer between us and reality – like splashing another coat of paint on a wall. The process 'fixes' the story so that it does not fade, but also it distances us from the event.

The cockpit area in the chalk produced part of a parachute harness, the canvas cover of an oxygen mask, a badly-bent compass and a gunsight. At the six-foot level was found part of the parachute ripcord, and below that to a depth of eight feet in the chalk a buckled spoon and more human bones. The spoon was curious, I thought. Found bent, and so deep in the chalk, it could hardly have been thrown in after the crash. Was it perhaps a mascot? Parts of a disintegrated Merlin engine were at this level also, with the crankshaft and propeller boss around eight to nine feet down. Distinct burn-mark stratification in the sides of the excavation marked precisely the upper lip of the crater caused by the impact. Compared to the intact Tudor bones and skulls I had been seeing under water recently, there was a great difference, for all here were both broken and abraded, apart from being unearthed in many levels. There was no doubt whatever that this Hurricane had dived in under full power at high speed. Had it gone in vertically?

The answer for me came when, after a mealbreak, I saw the discovery of the four missing guns. They were standing absolutely vertical in the soil, butts uppermost, two feet below the surface, muzzles about four feet down. The final proof was dramatically there: this Hurricane had indeed gone straight down vertically. Without doubt, it was the same crash my brother and I had witnessed in 1940 when I was twenty-two. It was sad that this pilot had died at twenty-three, but in that year I did not expect to live to twenty-three, nor did most of my friends. I had stolen from history and the politicians thirty-nine years of life to which I was not entitled. That was a pleasing thought. A kind of delayed 'after-action' reaction, when you feel like shouting: 'I'm alive! I'm alive! I'm alive!' because it's so extraordinary. This was that feeling, in a milder form.

I was not really sorry for the pilot. He died doing what at that time I longed to do. He must have been shot at once, fallen over the controls, roaring off 15,000 feet in a few seconds. If not quite dead when shot, then dead within seconds. Very quick, very clean. And on that sunny, warm morning, the stubble field was a very pleasant place in which to lie, looking away over the peaceful farmlands and woods for which he had fought. The Wealden Group had done his relatives a favour, I thought. They could not have known how swift was his end. How merciful.

There I was quite wrong. Extraordinary though it was that he had remained officially 'missing' and therefore unknown for thirty-nine years, what is even more extraordinary is that, so far as officialdom is concerned this pilot is still missing. His bones were not given ceremonial burial with relatives present, but quietly disposed of in a local incinerator. The guns, which an influential museum wanted for a display intended to depict the weapons used in the defence of Hampshire over the centuries, were refused and ordered to be broken up so quickly that there was no chance of obtaining a reprieve.

They had been able to do this because the Wealden Group had failed to come up with a number that officialdom was in a position to verify. The team had scraped and rubbed to bring up the numbers on the gun butts, successfully; but that was not enough. My evidence regarding the date of the crash (given to the police in an official statement on 6 October 1979) must have been disregarded also; for if you accepted that, then you only had to check to see how many Hurricanes were lost, and where, on 6 November 1940, and which pilots were still unaccounted for, to reach a certain answer. The Wealden Group had already done that, and in due course I too checked.

The news of their 'dig' produced a letter to the Portsmouth *Evening News* from yet another witness, a former cartoonist for that paper who from 1939 to 1942 had been Senior Gun Control Officer at 35 AA Brigade, Fareham. Conflicting reports of a fighter having been shot down were received, and so as to put in a claim for the 'guns' if possible, the officer went at once to the crash site. When he arrived, there was no one else there and the aircraft was still burning under the ground, with a small part of what looked like the tail protruding. An RAF salvage officer arrived from Tangmere and established from the number on the tail that the aircraft was a Hurricane from Tangmere, telling the artillery officer that this number would also reveal the identity of the pilot. Why at the time nothing further was done seems extraordinary. Also, there was indeed a Hurricane squadron at Tangmere and their Operations Record Book does show an entry for that day: 'Squadron patrol base; then Portsmouth A.25. Sgt. ____ missing.'

Helmut Wick and John Dundas were genuinely missing and remain so. Rudolf Pflanz, the last man to see Wick alive, was promoted Hauptmann, decorated with the Knight's Cross, scored fifty-two victories – nearly as many as Wick – and was killed in 1941, as was Wick's successor as Kommodore of JG 2, Wilhelm Balthasar.

After these personal experiences with the Widley Hurricane, the apparent complete disappearance from earth of Guynemer and his Spad in 1917 seems less puzzling; and I am now quite prepared to accept P.J. Carisella's extraordinary story regarding the abandoned bones of Manfred von Richthofen.

Chapter 8

THE GHOSTS OF WIGHT
1940–1941

It was 7 October 1979, the day after I had attended the Wealden Group's Hurricane dig. My wife and I were sitting by the fire in John Cleaver's farmhouse outside Ryde in the Isle of Wight. This time it was a ship and not an aircraft I had come to investigate. John was a member of the *Mary Rose* diving team and when a fisherman had told him that the seabed had shifted and exposed part of a wreck buried under the sands, which not even the local grandfathers knew about, he thought it might be worth our while to take a quick look. What we had seen that afternoon was not much – just two timbers emerging from the sand and weed, for the seabed had begun to cover it again – but worth a preliminary underwater excavation next year, we thought. Then we turned idly to other topics.

'What are you working on at the moment, Mac?' said John. 'Another book on the *Mary Rose*?'

'No, this one's to be about mysteries of the air, including the story of the Montrose Ghost.'

'Would you be interested in a local ghost?' John inquired. 'There's a neighbour of ours, a Mrs Brown, she's been seeing the ghost of an airman for years. More than four years, I think. She rang up my mother to see if we knew any local history because, she said, "I keep seeing this chap." She thought we might know who he was.'

When I was first told about the Montrose Ghost, I was a child; too young to believe or disbelieve, let alone probe the evidence. It was just an exciting story about someone my mother had known, almost a family legend. But in the short period before the war, 1933–1939, when I was flying light aeroplanes, I had two experiences in the air which were so strange as to be inexplicable; and during the war, two more such experiences, both on the ground, which included seeing an actual ghost. Doubtless there were other experiences which I have forgotten, but these four I never forgot. In spite of that, all were different, in that each required a separate category of explanation, if explanation there could be. At the time, they quite satisfied me as proof of a world beyond this one.

The first one occurred during an early cross-country solo in a Gipsy I Moth when I was about seventeen. I chose to fly down the coast from Portsmouth to where a girl friend lived near Bognor and to put on a display of elementary aerobatics, finishing with a spin. As the manoeuvres would be made over the sea, they would endanger no one. No one except me, that is, and I did not intend to take risks. I knew – or thought I knew – at exactly what height to put the stick forward and give opposite rudder in order to bring a Moth out of a spin with a height margin of 500 feet.

This particular Moth I had not flown before; it had a tendency to limp along with one wing low. I suppose it was badly rigged, and perhaps that explained the technical side of what happened when after a few dives and turns I pulled up to 2000 feet, hauled the nose in the air and kicked hard right rudder. The little biplane, really a small BE2 from which indeed it was derived, fell away on one wing into the spin with the nose down almost vertically, so that I was looking between the centre-section struts nearly straight down at Bognor beach covered by a few feet of water. The trick for recovery, in which I was well-practised, was to put the stick forward to gain flying speed while applying opposite rudder. I did this.

Nothing happened. I was still looking forward over the engine cowling at Bognor beach, as the fuselage went round and round slowly in the spin. Stick right forward. Hard left rudder on. Nothing more I could do.

It was then that I had the most remarkable, surprising sensation. I felt utterly detached from what was happening and also from what was about to happen. I felt absolutely calm, freed from my body and, poised above it, watching myself spin down to death.

My girl friend saw the Moth go down behind the houses lining the seafront, still spinning. The machine came out of the spin so much at the last moment that I could see right down through the water to the sand on the bottom.

Then I was alive, flying at wave-top level and back in the

cockpit, in control of the machine once more. I never forgot the calm sensation of perfect detachment from the body which I had experienced just that once, during the moments when I had done all that could be done and had only to wait. In a somewhat similar situation some years later, in a German sailplane, that feeling of aloofness from the body was not repeated. But I have heard of other people who also have experienced this peculiar splitting of one's self from one's body; it is probably not all that uncommon. How to explain it? Personally I felt that the phenomenon tended to support those religious beliefs which distinguish between the physical body and the imperishable spirit; between this world and the next.

Shortly afterwards, however, I had another experience in the air which was of an altogether different nature, not so readily explained. I was at 2000 feet over the Thames on a winter afternoon when I was caught very rapidly by an old-fashioned 'London particular', a very dense fog which would spread with dangerous speed over the rooftops. I was somewhere east of Tower Bridge when I noticed that East London was starting to disappear under this insidious grey wraith. If there was no way home to Portsmouth, my fuel margin being sufficient only for good weather, my only chance now was to get in to an aerodrome west of London. The only one I knew was Heston, so I pushed forward on the stick and dived towards the river. If you can't go over or round, you must go under; and I had no option. When I passed Westminster, I was at the same level as the windows of the Houses of Parliament.

Luckily, this time I was flying a Moth Major. More powerful than the Gipsy I, it had a better performance and, far more important in this context, it handled precisely and delicately. I had need of this, for the fog forced me down towards the water, below the level of the Thames bridges, so that I was compelled to jump over them, like riding a horse over so many fences. Showjumpers do not normally compete while holding the reins in one hand and a map in the other so as to predict their way round the course, but the equivalent I had to do, because this was the first time I had flown over London, and I did not know it well even from the ground.

One hand on the stick, the other holding my map, I recall jumping over one bridge which was on a bend of the river, so that I was turning, left wing low, as I went down the other side and found myself looking into the open mouths of a couple of anglers sitting in a boat. I don't claim to have counted their teeth, but the look of amazement on their faces would have been funny in normal circumstances; I was only twenty feet from them and my wingtips were much nearer.

Once on the straight again, I had to look out for some large gasholders which, according to the map, might be visible from the Thames as a guide to the whereabouts of the Great West Road (as it then was), which ran a mile or so south of Heston, if I judged it right. Momentarily, I had outflown the worst of the fog with the river as my guide, and now I was lucky enough and sharp enough to spot the gas holders. I changed course, left the Thames, and flew down the Great West Road.

The next transition was more difficult, for Heston aerodrome lay to the north with no landmarks to show me when I should turn north towards it. Aerodromes were small then, and did not present the enormous target of a modern jet terminal. Further, the combined effect of the fog and winter dusk was to blot out the horizon altogether. I could see nothing ahead over the engine cowling except vapour. There was no horizon. I could see land only by looking down below onto a small circle of visibility I carried with me, which was about a hundred feet across to start with, decreasing within minutes to fifty feet or less. There was no landmark at all. I would have no idea when Heston lay immediately to the north of me.

Then it happened. I received the command: 'Turn now.' I didn't exactly hear a voice. I certainly did not have a premonition, or a hunch. On the contrary, I was *told*, by something or someone outside me, that now was the time to make my turn.

I banked over to the right and was forced lower and lower by the fog and the darkness; within a minute, dead ahead between my centre-section struts, I saw the main hangar at Heston! I kept right down on the deck for my circuit, and apart from having to violently leapfrog the radio aerial which appeared between the masts, I made an uneventful circuit and landed in visibility of fifty feet or so, with not all that much petrol left in the tanks.

And it was all due entirely to the 'voice' which had told me to turn. It was *not* an audible voice, but it *did* come from outside me.

Certain mystics have talked about their 'Voices', which they interpret as instructions from God. Their opponents replied that it was the Devil speaking, while modern thought suspects that the poor things were deranged or suffering from sexual deprivation. My readings of history were changed by this experience, as it gave me an inkling of what might have been the real nature of these reported events.

But these two aviation experiences were strictly in the present. The next two were both on the ground, but while one hinted that you could sense the past, the other implied that one could foresee the future. The first was a dream, I think, rather than a presentiment. I found myself walking along Burgoyne Road in Southsea, and looking east saw that several houses showed empty roofs, having been burnt out. Just those houses were to be burnt out in an air raid two months later. Perhaps it was coincidence, but if not, the implications are disturbing. The thought that the future, or part of it, might be fixed is not pleasant.

It was during the war that I actually saw a 'ghost' – just the one, and just once. The occurrence took place in a famous Scottish castle where I was sightseeing on my own and where my lack of previous detailed knowledge was not remedied by the labels in the various rooms, for they all seemed to read 'Visitors are requested not to touch'. I 'saw' a girl in a flowing red dress with perhaps white or gold upon it hurry through a room towards the far door (which, I subsequently discovered, led to a private staircase leading to a tiny, discreet dining room). I could make out the girl quite clearly, and the pattern and colours of her dress. But never for one moment did I think that an actual person was there before me. The figure was only partly in the room; it was also partly in my mind. So you could say that I could 'see through it', as with the traditional wraith-like ghost. Much more substantial than the vision itself, however, were her emotions. She was happy, excited; she was going to meet a lover. These feelings of hers came over to me instantly and with full force. I understood them at once.

Afterwards, I found that there was an historically exact basis for what I had 'seen' and 'understood'. There was a record of such a girl, including a representation of the dress I had seen her wearing; and the private staircase and dining room had often been used by her. The only discrepancy lay in her height: the real girl had been taller than the 'ghost' I had seen.

That incident convinced me that one could indeed sense the

past, and I thought that perhaps if an emotion, hate or love, was strong enough, it might go on forever in the place where it had been first experienced. This seemed to me a proposition not very difficult to accept, much easier in fact than any foretelling of the future or outside 'voices', which I found hard to rationalise.

Especially with 'ghosts', seeing really is believing; indeed, I am rather ashamed to admit that if a story of the apparently supernatural comes into one of the four categories which are all I have personally experienced, I am prepared to consider it, whereas if it is in a different category strange to me, I approach it with suspicion. I recognise that that is arrogance based on ignorance, but there it is; at least I proceed from a known base. Consequently I thought that John Cleaver's story of the ghostly 'Pilot' who had begun to appear to his neighbour Mrs Brown, would be well worth while investigating at source.

So, soon after our chat with the Cleavers, my wife and I again went over to the Isle of Wight, this time to see Mrs Violet Brown, her husband Hubert ('Hugh') and Garry, one of their sons. Their house at Binstead near Ryde had been built about 1954 on what had been pasture land during the war, bisected only by a road leading to Newnham Farm where the Cleavers lived. Violet and Hugh showed us into the drawing room and left us for a few minutes to prepare tea and sandwiches. Off this room there was a sunlounge. Instead of sitting down, my wife at once walked through into the sunlounge and stood to the left of the central window, looking into the garden. After a short hesitation, for I felt we both ought to sit down in the drawing room and wait there, I also walked through into the sunlounge, moving to the right of the window. Between my wife and me now was the central pane of the window overlooking the garden.

After tea, I expected to be invited over to the old house in which the Browns had lived before and where I assumed the ghost had been seen. This proved to be my misunderstanding, for I learnt now that the 'Pilot' had first appeared at this house and only behind that central window pane in the sunlounge. I had not sensed anything at all, and I don't think my wife had either, but we had indeed walked directly to that window and stood on either side of the central pane, although ordinary politeness suggested that we should sit down in the drawing room. I was a trifle embarrassed that we had walked through in such a possessive way.

Then Violet and Hugh came in with the tea and sandwiches, and we sat down to talk, facing the sliding glass door which led to the sunlounge. Violet explained it had started about three or four years before in late 1975 or early 1976, always in the evenings around half past nine to half past ten. First, there was what she called a 'loud quietness'. Next, she heard the buzzing sound of a bomber flying over. Then she felt compelled to look out towards the sunlounge. The window was steamed up. 'There stood a chap. Six feet tall, curly hair. Wearing bomber jacket and dark tie.' Violet could make out no more than that, although he stood there for fifteen minutes and then disappeared. 'I had these experiences when I was low,' she explained. 'when I was exhausted physically and mentally.'

Remembering my own single experience of seeing a ghost, I asked if it was solid. Violet replied that she was convinced that it was someone, but at the same time knew it wasn't substantial. I hadn't told her of my experience, but that exactly described my own impression, so we were talking about the same sort of thing. I then asked if Violet had been able to feel the ghost's emotions. She replied: 'No, except that he wanted to convey a message of some kind.' I asked why she believed it to be a bomber pilot. Violet said that she had *thought* Pilot and *heard* a bomber. There was no insignia on the jacket, which was the sort aircrew wore. Then she added, unasked: 'What struck me most was his face. The curly hair was distinct, but the face was blank, there was no expression at all.'

Violet explained that she could sense evil, or good. She could also sense perfume when there wasn't any – for example, flowers when there were no flowers, cigar smoke where there were no cigars. So there was a history of previous experience, although not of the same type. Violet had not told anyone about the appearance of the 'Pilot', but her daughter Lynda was psychic, and one night had been sitting in the room with her and the dog. 'Mummy,' she asked, 'is that a plane?'

'Yes.'

'But it's not a modern one, is it?'

'No, it's a bomber,' replied her mother.

The dog stirred uneasily, then Lynda looked at the window and said: 'There's someone out there. Can you see it, mummy?'

The dog began to bark, and at the same moment her husband came in and as it was late, put on the light. Then he walked right through the 'Pilot'. He couldn't see him at all.

Lynda had been upset by the experience, but Frank, her

husband-to-be, did not believe the story. He thought it was just female imaginativeness. Soon after, he came to visit them and was left in the drawing room alone. He was not there long! Chalk-white, perspiration running down his face, he bolted into the kitchen. 'I'll never disbelieve them again!' he said.

Later, when Frank was leaving in his car, he felt something get into the vehicle with him. He took it to Quarr (almost gave it a lift, so to speak), and did not see the thing after that. So the Brown family thought: was there perhaps a plane crash by Quarr? Mrs Brown telephoned Mrs Cleaver for facts of any plane crashes nearby, because she felt that the pilot was trying to get through, and if one found out who he was, one might be able to help. When Mrs Cleaver was told that the 'Pilot' was tall, dark, with curly hair, she said: 'That description fits Mr Plant's son.' This was a neighbour whose son had been killed in the RAF and who had since died.

Mr 'Hugh' Brown was often away from the Island, serving as Chief Refrigeration Engineer in Cunard ships carrying fruit cargoes. Often, when he came home, he'd be told: 'The Pilot has been back.' He himself was occasionally able to sense the 'Pilot', but not so often as his wife. Once, early on, Lynda had said to him: 'There's someone with us, Dad. Don't be alarmed.' Even when he couldn't see the 'Pilot', Hugh became aware of a definite sequence associated with his being seen. First, he said, the room went very quiet and there was a draught – although no windows were open. The clocks took over with the quietness, and the feeling in the room became cold – not the room temperature, he stressed: that remained warm. It was a *feeling* of cold, not actual cold.

When I asked Hugh my standard questions, based on my own solitary experience, he replied that the 'Pilot' was solid, except for his head – which was obscure. 'Foggy' was the word Hugh used to describe its face, just as Violet had said 'Blank'. Hugh saw him with a flying helmet on, Violet without. They agreed on the leather jacket. When I asked if he could sense any emotion being felt by the pilot, Hugh said no, but that they both thought he was looking for something, which they hadn't yet been able to convey to him.

I then asked questions regarding the noise of the aeroplane. Hugh knew the sounds made by wartime aircraft, British and German, just as well as my wife and I did. He said it was a bomber, not a Spitfire or Hurricane. It was definitely not German. It was a twin-motor job, like a Wellington or Whitley, not a Lanc. It flew over from north to south just west of the house.

A third member of the family came in when we were talking. This was their son Garry who, like his father, had served for a time in the Navy. Garry had never seen the ghost nor ever heard the bomber. Not once. But he had previously felt strange presences in a ship and also in a church, and in the house where we now were, of his grandmother. That was all. For his part, Hugh had an occasional horrible feeling of an evil being following him around the house – when the 'Pilot' wasn't there. He had felt the same thing at sea, when there were deaths in ships. At one localised point on the Isle of Wight, a place called Dame Anthony's Common, not too far away, he had also experienced this sensation of evil.

Trying to save time, I was now no longer writing complete sentences; I was just jotting down the gist of what the speaker said. I summarised Hugh's impression of the Common as 'Bad feeling. Murder'. I did not see what the evil atmosphere of Dame Anthony's Common had to do with the 'Pilot' of Binstead, who seemed to be benign.

Both the Cleavers and the Browns had programmed me towards aircraft crashes instead: the Cleavers towards two Hurricanes which had collided over Binstead Lodge Farm, which had belonged at the time to John Cleaver's father; the Browns towards a twin-engined machine, the bomber which they heard when the 'Pilot' appeared. Of course, I couldn't probe every crash on the Wight during the war. I decided to follow the Cleavers' lead and start with the Hurricane collision, because it had happened close to where the Browns' new house now stood, on land that had been pasture during the war. Tacitly, I accepted their assumption that what we were looking for as a possible candidate for the identity of the 'Pilot' was someone who had died in such a terrible way that his soul could not rest.

Of course, there were two Hurricanes involved, so two crashes and two crash sites to identify. The period appeared to have been about the end of November 1941 – people were vague about that. The general impression was that the two machines had been called up from Tangmere and had then been recalled; that one pilot had got the message and turned back, and that the other had not; and so they collided. However that may be, some people could still remember the bang as the two Hurricanes touched in mid-air. One machine crashed by Binstead Lodge Farm, while the pilot parachuted into a field called Eleven Acres. Witnesses could not be sure of his accent: one thought he was an Australian, others Canadian. Mrs Winfrid Drudge recalled that he was found wandering along, shouting 'Where is my buddy?' Then he was taken into the farm to wait for an ambulance. Miss Vera Thorne, then about thirteen years old and living there, remembered the farmhouse swarming with air raid wardens, police and neighbours, but that this pilot was not badly injured. The Head Air Raid Warden of Ryde, Rex Burton, was one of these people. 'Judging by his accent, he was Canadian. He was fairly all right, more concerned for his colleague than for himself. I told him the other pilot was "All right" (actually, he was in the mortuary, but this pilot was in shock and I didn't want to make it worse).'

While this pilot, surrounded by sympathetic onlookers, was mourning his friend, the other pilot still lay undiscovered on Dame Anthony's Common in the wreckage of his machine, because the nearest witnesses dared not go there. They were the occupants of St Wilfrid's Cottage in the village of Haylands, not far from Binstead. There were two ladies, Amy Saunders and Irene Cheverton, with their five-year-old children, Robin Saunders and Valerie Cheverton. Both their husbands were away in the Army, and the women did not want to venture out into the night, not knowing who was in the plane, which might have been German. It had crashed into a field only about 150 yards away, having missed the cottage by not very much and scaring the occupants. Because of the blackout regulations, they had first to turn off all the lights and put the cottage in darkness before they were able to open a window or door and look out. What they saw were the lights of torches in the dusk, as a search party from Binstead Lodge Farm came down towards the crash.

'Eventually men brought the body of the pilot up from the field and placed him in the garage near our cottage,' recalled Mrs Saunders. 'Soldiers were placed on guard there and by the crashed Hurricane. I didn't see the plane that night, but the crash site was visible for a long time, although now there is no sign. The cattle have trodden it in.'

So there was one man who had died violently on Dame Anthony's Common. But this crash occurred a little distance from Binstead. Perhaps the pilot who had parachuted safely over

Binstead had been killed later and came back looking for the friend he had inadvertantly sent to his death? Was that a wild suggestion? No wilder than the fact that although Amy Saunders had never heard of any ghost at Binstead, she did know that an apparition had been seen on Dame Anthony's Common in about 1960 – before the 'Pilot' first appeared to the Browns.

'I haven't heard about the ghost of an airman seen around the Binstead area,' she said. 'However, my youngest son David, born in 1946, came home at dusk with a school friend, Robert Hendry, and they said they had seen, they thought, a ghostly figure gliding through a hedge. This hedge divided Dame Anthony's Common from the field next to the one where the plane crashed, killing the pilot. I'm afraid we were a bit sceptical at the time, but the boys were quite serious they had seen this figure. Of course, this incident happened around twenty years ago now and my son is living in Australia, having joined the Royal Australian Air Force there.'

David Saunders had been, he thought, about nine to twelve years old at the time. In 1980, when he wrote to me, he was thirty-three, and enclosed a sketch so I could see where it had happened. He and his friend Robert Hendry had been playing on a bridge across a stream, the same stream beside which the Hurricane had crashed. One way, the road from the bridge led to Ryde, the other way via a field path to the village of Havenstreet. There was a boundary hedge running at right-angles to the stream, which divided the field from Dame Anthony's Common. The time was dusk but in the summer months, and therefore much later than on the winter day when the Hurricane pilot died. David and Robert were still playing on the little bridge when they noticed this 'white form', not walking but drifting from the direction of the Common, through the hedge and into the field with the pathway.

David wrote from Queensland, Australia: 'To me the form did not seem "real", i.e., as an earthly being does. We both sensed it was something different and after a few moments we scurried to my home at Sallers Road, Haylands. I believe we described it to my mother, father, and my brother Robin.' Amy Saunders, having read her son's letter, commented: 'All I remember is how certain they had been at seeing this ghostly figure floating through the hedge.'

This clearly was not the Binstead ghost, although seen close to the Hurricane crash site. Then from another source, a Mr A.J. Atkins of Ryde, who then knew nothing of the 'drifting' ghost by the stream, I received news of either this ghost or another in the same area, also seen by a schoolboy:

> Some eighteen years ago, while still at school, I was informed by a classmate that Dame Anthony's Common was haunted by the spectre of a German pilot. The Common lies just south of the recent development on Binstead Lodge Farm. I have never met anyone else with any knowledge of the phantom's existence. My classmate also mentioned, possibly as an embellishment of his own, that the ghost was the spirit of a crewman who survived the actual crash but was shot dead by trigger-happy soldiers, fresh from Dunkirk, as he staggered from the wreckage. He also stated that the aircraft was either an Me 110 or a Stuka (I forget which but it carried a crew of two) and that his mother witnessed the incident. The story is possibly the bloodthirsty fantasy of a schoolboy, but as I have lost contact with him for some years now, I have been unable to probe further.

Bloodthirsty it was, but no fantasy. Half the people who had witnessed the collision of the two Hurricanes had either seen or heard of the Me 110; many more again knew about the Messerschmitt only. Again and again, with only minor differences of detail, I heard that the 110, harassed by fighters, had made a crash landing at West Ashey, some miles from Dame Anthony's Common, and that one of the two occupants had been shot by Channel Island Militia and his body left to rot.

Oscar H. Smith was working at Nunwell Farm when the big two-seater came in from the sea, skirting the high ground between Bembridge and Brading. 'I can well remember this plane approaching Nunwell Farm from the east, flying rather low and in trouble. It passed close to me, about a hundred yards from our house and just above the trees in the meadow adjacent to the farm. I thought it could only get as far as the west side of the farm but in fact it came down at West Ashey half a mile away.'

Mr F.T. Aylett lived in the farm adjoining West Ashey. He saw two British fighters round the German plane and heard a lot of machinegun fire. Brian Warne, then aged twelve, saw it too from a couple of fields away. The Messerschmitt crashed into a hedge bordering a road, bounced clean across the road, struck a telegraph pole with its left wing and finished in the cornfield beyond, undercarriage up and apparently not too badly damaged.

What happened then, after the Messerschmitt had come to rest inside the cornfield, was not witnessed directly by any of these three people, but they were close enough to learn later at second or third hand, and their stories match. Oscar H. Smith said: 'A few minutes after, Lady Oglander came down the road in her motor at speed and I noticed as she passed she was wearing Red Cross uniform as she was Commander of the IW Division. I decided not to get my motor to see where the plane was. We were told not to and Lady Oglander's husband, General Oglander, was head of the Home Guard and I was a member; also we were only allowed petrol for agricultural purposes.' The Oglander family had been a power in Wight for generations and had played a prominent part in Henry VIII's time in organising the Militia to drive out the French army which had landed on the island. However, next morning Mr Smith got Lady Oglander's story from the Bailiff. When she had reached West Ashey the Jersey soldiers who were stationed there had just shot the German, and when Lady Oglander asked; 'Why did you do that?' the corporal in charge had replied: 'He tried to fire the plane, Madam.'

Oscar Smith commented: 'I always understood there were several crew and the grapevine said they came out with their hands up and the Jersey boys shot them. All sorts of tales got about in those days, and this may not be true.' But Brian Warne, aged twelve, heard the same story. 'The Jersey and Guernsey Regiment were billeted 200 yards away at West Ashey Farm. One German got out. The other German was still in the plane, fiddling about. They thought he was going to fire it, so they shot him. Practically blew his head off. I don't know how many fired. I did not see this – it was the story they told. Of course, Jersey and Guernsey were occupied then.'

Mr F.T. Aylett, who farmed Roylands, next to West Ashey, told the same tale: 'One was taken prisoner – the gunner? One was shot in the cockpit – the pilot? By Channel Islanders, Guernsey and Jersey Militia – evacuees who had been conscripted. It was very hot weather. They left the dead body there for at least a week.'

It proved impossible to find the grave or to get a direct witness of the shooting, but Reg. C. Eve, now seventy-three and living in Jersey, had the job of moving the corpse:

> The unit concerned was a platoon of the Royal Jersey Militia, of which I was a member; later the whole unit joined the Royal Hampshire Regiment and became the 11th Battalion. At the time of the crash of the Me 110, I happened to be the ration-lorry driver which called at the farm some hours after . . . and I was informed that the rear gunner had been taken prisoner, and the pilot who was dead was still in the plane. I forgot the incident but a day or two later I was ordered to take the lorry and a party of four to Ashey to take the pilot's body from the cockpit of the plane and deliver it to a sexton or gravedigger at Ryde Cemetery, and this was done with reverence and care.

So the story was true, the 'grapevine' vindicated (give or take a little uncertainty among the witnesses as to whether it was the pilot or the gunner who was shot and whether it was Guernsey or Jersey men who did it). Therefore there were two dead pilots in the area, one British, one German, who if our superstitious assumptions have any basis, could not be expected to rest easily. And one indeed was the pilot of a twin-engined bomber, for I was able to prove that the Me 110 was the fighter-bomber version of this two-seater. The coincidence involved was not so much remarkable as completely astounding. For it turned out that I had seen, described and sketched that particular machine, that identical Me 110, six weeks or so after the crash, and had even possessed a part of it (since lost).

I suspected this early on when Mrs Winifred Drudge, after describing the collision of the two Hurricanes above her head, went on to talk about the German plane she saw hedge-hopping towards Ashey Down, which was afterwards put on show at a local cinema where people paid sixpence to see it. I looked up my Battle of Britain diary and found that on 28 September I had seen on show in a Southampton shop window (Messrs Edwin Jones) a Messerschmitt 110. My 1940 notes read:

> Most remarkable was the wingspan. It was the biggest two-seater I have ever seen. It had been shot down in flames by a Hurricane on Ashey Common, I.O.W. There was a squadron device – something red on a white background – on the nose, but this had been too badly damaged in the crash to make out exactly what it was. The starboard engine mounting showed traces of fire. A navigating instrument was on view, likewise the pilot's boots, and his flying suit – with a bullet hole in the left shoulder.

I never thought for a moment that the shot might have been fired while he was on the ground. But I did buy for five shillings part of the fuselage with half a black cross on it and four bullet holes in it. And for good measure I did a watercolour sketch of the machine including what little I could make out of the unit insignia. Forty years later, it was enough.

The machine had come from a Luftwaffe experimental unit, *Eprobungsgruppe* 210, which possessed two squadrons of Me 110s and one squadron of Me 109s, all of them converted to fighter-bombers. I had seen them in action at least three times, twice when they were attacking the Spitfire works at Southampton, once when on their way to flatten Cunliffe Owen, another aircraft factory.

Although this information is satisfactory in its way, some major questions still remain mysterious. First, who was – is – the 'Pilot' of Binstead? The Browns thought of him as British, the pilot of a British twin-motor bomber, perhaps a Wellington. Could he be a German airman, the pilot of a twin-engined fighter-bomber, shot dead when he must have thought himself safe? If so, why did he appear so long afterwards? What did he want? Or was he instead the wraith which the two schoolboys saw 'drifting' through the hedge on Dame Anthony's Common? It is probable that no one will ever be able to answer these questions.

But perhaps more fundamental is to wonder why we ask these questions at all. Why is it that we seem to demand a logical explanation for the appearance of a ghost, and do not feel happy until we have discovered or manufactured one? The 'Montrose Ghost' was a classic in this respect, because C.G. Grey was able to introduce convincing explanations for the timing of both the appearance and eventual disappearance of the apparition.

Nevertheless a similar type of story was to become associated with another RAF base, Lindholme near Doncaster, which operated in the Second World War. Witnesses ranging from airmen to Air Commodore, and including a doctor, reported seeing an apparition. The sightings began after the war, at intervals of about a year, and took the same course in each case. The witness was approached by a man in mud-stained clothing who asked the way to some part of the base – the sick quarters, or the officers mess, or the operations room – and they vanished into the darkness.

Then in the 1970s a marshy area near the aerodrome, known as Lindholme Bog, was drained and thirty feet down in the peat a wrecked Wellington bomber was found, with the crew still in it and unidentifiable. After they had been dug up and reburied in the grounds of Lindholme church, the hauntings ceased. The men were thought to have come from a Polish Wellington which, returning from operations with only the pilot still alive, although badly wounded, had gone down into the bog 300 yards short of the aerodrome's main runway; and had very quickly sunk from sight.

Chapter 9

NO CHRISTMAS IN CONWAY STREET
Portsmouth: 23/24 December 1940

The moan of the siren might mean nothing at all – a raid heading somewhere else or a reconnaissance machine – or it could mean death very quickly. After eight months of real war from the air the inhabitants of Portsmouth and other south coast towns had become expert at telling the difference, and at reading the various noises which indicated what sort of death was coming at you and how near it was likely to be. But this night everyone was to be baffled. The sound and its results were totally strange and extraordinary, and have never been explained. Every few years the local Portsmouth *News* prints another selection of letters asking what happened that Christmas night long ago, and no official reply is ever received. Like an unsolved crime, 'the incident was never closed', as one reader put it.

The blood-chilling wail which once heard in earnest is never quite forgotten sounded once again at 18.16 hours – 6.16 pm by peacetime clocks – on 23 December 1940. It was a very dark night, the sky nine-tenths obscured by cloud. This gave little protection to Portsmouth, which is virtually an island, surrounded by water on three sides and a canal and military moat to the north. Water shows up well from the air, even at night, and the distinctively tidy fretwork pattern made by the tidal basins and docks of the naval harbour on the western side of the city give valuable reference points to any bomb-aimer. To the east there are two harbours, Langstone and Chichester, separated by Hayling Island, which make good check points from the air. I did some night flying from Portsmouth airport before the war and was surprised to find how easy it was to orientate oneself.

Most of the witnesses, though not all, testified that the plane came at them from the east. Frederick Bishop was a Royal Marine manning a Vickers heavy machinegun on the south side of Eastney

Barracks, facing the sea. He heard the plane cross the coast at Selsey in Sussex and then turn west parallel to the beaches for its run-up towards Portsmouth. It was fired on by batteries at Hayling, Langstone and Portsdown Hill, but the throb of the engines did not alter. It passed slightly on the north of the barracks and then, when it was northwest of him, Bishop heard 'a horrible sound as if a piano was dropping with wind whistling through the wires'. There was no change in the engine note, but the horrible sound continued until there was 'one hell of a whitish flash as it hit'. The Guildhall, Semaphore Tower and the Dockyard cranes were outlined by the flash. These are all in on near the Naval dockyard, the largest in what was then the British Empire.

The plane then turned towards the Isle of Wight on a southerly course and as it went over the eastern tip of the Island, Bishop saw it come down in flames, with the guns firing at it. Later, the body of a German wearing a parachute was washed up on the slipway by Eastney Barracks; the corpse had no flesh on its face or hands. Bishop thought this might have had something to do with the incident. The extraordinary noise before the explosion puzzled him; he conjectured that it might have been made by parachute shrouds entangled with a falling bomb.

Another lookout on duty that night was Frederick James Kistle of the Auxiliary Fire Service (AFS). He was on top of an observation tower at the Eastern Road depot, about four miles north of Eastney Barracks. He also heard the plane come in from seaward and the guns of Hayling or Fort Cumberland begin blasting away at it over the City. Alarmed, he shouted down to the men on the ground below him: 'The so-and-sos are firing at it directly overhead!' 'No sooner had I said this,' he recalled, 'when there was a flash and an explosion in the sky and the plane started to fall and a few seconds later there was an almighty explosion. One of the engines I believe was found inside the dockyard wall. It seemed fairly obvious, by the size of the crater, that it had come down with its full load of bombs.'

Mr L.R. Stubbington, living then in central Portsmouth, made much the same remark to his mother, although more decorously:

> The familiar throbbing drone of a German bomber approaching, considerably louder than normal, was heard. Either lower or closer than usual. At this point the ack-ack guns opened fire with several sharp cracks. They moved mobile guns around the

town, and these sounded very close. I remarked to my mother: 'What a time to open up, with it right overhead.' At this point the aircraft noise increased, with a swishing noise increasing to end in a violent thump which could be felt through the floorboards. The only damage we suffered was one cracked window.

His mother's house was about a mile away from the impact.

William Wills was living by Baffins Pond, two miles from the impact point and under the route of the bomber.

It was early evening when a terrific rushing and displacement of air was heard approaching from the east and passing overhead. A few moments later a violent explosion occurred and the house trembled. From experience of many later incidents, neither landmine nor heavy bomb caused such a displacement of air. Like many more people, I would like to hear the truth of this occurrence.

Police Constable Kerby, on duty at Fratton Police Station, also described it as a 'rushing sound'. Windows were blown out around him, and at first he thought the bomb had landed in Fratton Road, whereas it had impacted one and a half miles away. Roland Walker, aged thirteen at the time, heard 'a rush of air like a bomb, not like an aircraft in death throes in a throttle dive. A whooshing sound.' He thought the aircraft had cut its engines. What puzzled him was that no bomb fragments seemed to have been found, nor fragments of parachute, if it had been a landmine.

Two miles to the south, Mrs Eileen Skeates (then Miss Garner), serving in the AFS, heard the plane, and a local heavy gun known as 'Big Bertha' give a single blast; she felt a pressure wave, heard an 'ugh-ugh' sound from the plane, and a man shouted: 'He's hit it, he's coming down!' Then there was a violent explosion.

The explosion took place at about quarter to seven, during a change of shifts at the dockyard. Walter Jeakes was cycling home and had reached North End, more than a mile away, when he heard:

an aircraft and this extraordinary noise. I pulled up on my bike to listen to it. It was like somebody had dropped sheets of corrugated iron. A rushing, tearing noise like iron turning over and over, which went on for a long time; then a God-Almighty bang. I'd heard many bombs before, and this was nothing like them. My idea was, that it was an aircraft shot up and turning over and over, but a friend in the dockyard, Mr Thoroughgood, was on the investigation team, and he told me they found no aircraft bits.

C.M. Jeram was the leader of an eight-man trailer pump crew in the dockyard. Their action station was a brick-built shelter keyed in to the dockyard wall not far from the Unicorn Gate.

> There was a lull in the general noise of an air raid, and I heard a loud twittering noise, very much like the wingbeat of a sparrow, passing close, but on a slightly lower beat. Immediately the fluttering ceased, the whole shelter filled with thick, blinding dust. It was as though two giants with huge brooms had started to shovel all the ground dirt, dust, leaves, pieces of wood, etc., etc. into the shelter at each end. We couldn't see or breathe. We were all lifted up six or nine inches in the air, and landed on the floor (except for one man who landed on a plank and hurt himself). I must stress that I heard no explosion of any sort, only the loud twittering. We picked ourselves up and literally staggered outside and in the far distance (or so it seemed) I could hear crying and screaming.

Mr Jeram lifted the telephone and told his HQ: 'I think something has happened outside. I hear screaming.' They replied: Stay where you are. Your job is to look after the Yard, not outside.

So he went to check the other shelter where the trailer pump was housed, which had a steel blast door. The steel door had been blown in and had pushed the pump into the dockyard wall. There was other damage nearby, and the corrugated iron roof of the torpedo store was hanging off.

Reginald Cuss, who was a fitter in the torpedo shop, was cycling to work when he heard a swishing sound. Thinking it was a landmine which would shower the streets with broken slates from the roofs, he got down in the road and placed the bike on top of himself to deflect at least some of the slates. He got up unhurt and cycled on into Commercial Road, a main shopping centre about half a mile from the dockyard wall in front of the torpedo depot. All the windows were out, and broken plate glass littered the street; people were rushing to and fro, trying to help those who had been cut by the flying glass. When he got into the dockyard he found the rescue party helping dayshift men who had been caught inside the torpedo depot. In all, twelve men had been seriously injured, thirty-four slightly. Until the roof could be repaired, production of torpedoes would stop.

Frederick James Pepper, the conductor of a double-decker bus, had just been about to ring the bell for the driver to start up and leave the dockyard Main Gate, when he heard the noise. The torpedo shop was about $1\frac{1}{4}$ miles away to the north of where his bus had been waiting. A girl and a boy had just jumped onto the platform as he was about to ring the bell. To Mr Pepper, the rumbling noise to the north was 'like a train rushing into a tunnel'. Not like an ordinary bomb at all. The girl grabbed hold of her boy friend. 'Oh, it's a bomb!' she cried. 'No, it's something worse than that,' said Mr Pepper.

Then he rang the bell and the bus left the dockyard on its normal route to Tangier Road. At RN Barracks, the first stop, there was nothing unusual, but at the Commercial Road end of Edinburgh Road all the blackout material of the Colosseum Theatre had been blown out and the theatre lights were streaming across the road like searchlight beams. Going on through Arundel Street, every shop had had its windows blown out and the bus's wheels were crunching on broken glass.

All this, like the damage to the torpedo depot, represented merely minor disturbance around the perimeter.

Donald McGregor, BEM, who at the time was a section officer in the AFS and responsible for three fire stations, heard the sound from the Commercial Road station. 'There was no whistling noise as we had from falling HE bombs, but more of a roar like a train hurtling through a tunnel.' When he reached the scene, McGregor was 'appalled at the immense area involved from the blast and fire. Even underground gas mains were alight.' But some people never heard a sound.

Alan David Dart was twelve years old at the time. His father, Alfred Dart, owned two shops in Charlotte Street, which lay between Commercial Road and the dockyard wall in the area of the torpedo depot. He used to work late, printing a menu for Woolworths so that it could be delivered by 9 o'clock the next morning. That night, David went to meet his father and found that the girls in neighbouring shops were also working late, because it was Christmas, so behind the blackout curtains all their lights were on. Alan's father was not pleased to see him. 'You're a naughty boy, why did you come down?'

'I'd thought I'd wait for you, Dad,' said Alan and hung around for about a quarter of an hour while his father worked the printing machine, which was directly below a blacked out glass roof. Alan heard nothing at all. He merely saw a tremendous flash, of an 'orangey-blue colour', through the glass roof.

Then he was outside the building and in the street.

'I don't know if I was lifted there, or if I ran there. One moment I was with father by the machine, next moment I was outside in the street.'

And the whole of Charlotte Street was a blaze of light, because people had been working with the lights on and the blackouts up; and now not a shred of glass or a bit of blackout material was left in place.

There was a shout of 'Gas!'

Alan looked down Charlotte Street towards Conway Street and the dockyard wall and saw a cloud of what seemed to be gas. In fact, it was very fine dust coming from the ruined buildings and the broken bricks over there.

Then his father reappeared, injured. A large piece of timber had come through the glass roof, quite unnoticed by Alan, and hit his father on the shoulder. The two walked home together, and then Alfred Dart returned to his shop.

Elsie Godding (now Mrs Pitman) was aged eighteen at the time.

> Mum and Dad and I ran the 'Unicorn' public house outside the dockyard. Dad was called up as a ship's fitter, I also was called up and became an electrical spot welder in the dockyard. We three worked solidly from 6 in the morning to midnight, Dad and I in the yard and helping Mum to run the 'Unicorn'. So there was not much time for pleasure. The day before Christmas Eve our mince pies were made, the chicken stuffed ready for Christmas, a dinner we were destined never to eat; it was blown sky-high in the biggest explosion I've ever heard in my life.
>
> We were busy serving the night shift, due to clock on at 7 pm. The bar was beginning to fill up, the siren went, the double-red went almost immediately. We could hear this from the Naval Barracks (which was never very good for our nerves), the guns were blasting, a plane was overhead, then suddenly crash, all the lights went out, the windows came in, and we were all knocked unconscious.

I came to with my mother standing over me, and I thought I was dying as I was covered in a thick, sticky liquid. It was bottles and bottles of stout that had crashed all over me – not blood as my Mum at first thought. I don't really know what happened to the other people in the bar – I believe one sailor was quite seriously injured.

Mum and I staggered to the door and out into the street; the sight was like Dante's Inferno. Bodies, twisted bikes and buses, fallen lamp-posts, telegraph wires, smoke and dust, and the smell of devastation. We climbed over all of the many obstacles and made a slow, horrifying way to the Royal Hospital, keeping alongside the dockyard wall, gripping it for some support as the debris was mountainous. The Hospital was full of casualties, but those of us that were lucky enough to have survived the tremendous explosion were treated with great kindness and gentleness for our cuts and shock.

In fact, Elsie's spine had been damaged by the fall and later she had to spend a year in hospital. But that night, people remember a hospital queue for emergency treatment which was four deep and stretched for half a mile through the darkness and devastation. Miss Caroline Barrington, an ambulance driver, recalled: 'Then around a bend in the road came a crowd of people just like a football crowd. They were the walking injured.'

Edith Whitelock (now Mrs Sawyer) was twelve years old, one of a large family living in Trafalgar Street, which led to Conway Street. There were five of them down in the basement, which was steel-reinforced, with a steel escape hatch to the back garden and two bunk beds. Besides Edith there were the twins Aubrey and Kenneth, both thirteen years old; their mother, Mrs Louise Whitelock; and Mrs Baker from next door. The father was out firewatching on a coal barge in the harbour and seventeen-year-old Lionel Whitelock, who worked in the dockyard (we shall meet him later), had been sent to the shelter under Semaphore Tower.

The sirens sounded but as at first there were no planes, no guns, and the children started to pester their mother for permission to leave the basement shelter and go upstairs to get some games – Ludo, chequers, draughts, snakes and ladders. The twins lived on the third floor, the highest of all, while Edith's bedroom was on the second floor, under them. The children all reached their rooms and began getting out the games, when the twins, who knew the noise made by a German bomber, shouted out: 'Jerry's coming over again!' And they all rushed down the stairs.

Edith nearly got to the ground floor. The twins reached the second floor landing. Their mother came up from the basement to fetch the children down and had reached the top of the stairs. Then there was 'an almighty explosion'.

The top story collapsed, the debris fell through the house, and the lights went out. 'It was a shattering experience – everything falling about you. Our first reaction was to crouch down.' The children started to cough from the dust. Then they thought of their mother. The back door of the house had fallen inward on top of her, pinning her underneath; she was black and blue all down one side from the impact. As soon as they realised their mother had been hurt, the boys pulled the door off her. Then they lit candles and paraffin lamps, and there was light again. 'It was all too sudden for us to be really frightened', said Mrs Sawyer. 'We were much more frightened later in the Portsmouth fire blitz.'

Mrs Lilian Owen was a widow living in Chance Street, four roads away from Conway Street. Her mother, Mrs Emma Jane Harvey, who was crippled, lived there also, with Lilian's young brother, fourteen-year-old Cyril Harvey. There was also her own son, Edward Harry Owen, two years old. They were an Army family and Lilian's husband had been among those who had 'held the road to Calais earlier that year so that the others could get away at Dunkirk.' He was officially missing, believed killed, and she now received a widow's pension. It was not until 1941 that she heard that he had been wounded and taken prisoner and was in Stalag VIIIB.

Even so, they intended to have a family Christmas. Mrs Harvey was busy plucking a chicken which she had raised from a day-old chick to a fine eight pounds. Lilian's sister, who lived at Cosham with her two children, a boy and a girl, was due to arrive on the evening of 23 December, but as she was late Lilian went outside to see if she was coming. She stood waiting by the corner of Chance Street and Thomas Street, which was marked by a brewery stables, a strong building where the big dray horses were kept. I waited an hour on the corner for my sister, but she didn't turn up, recollected Mrs Owen.

I heard a dull drone, like a big wasp in your head. I thought: 'That's not one of ours.'

I didn't know where to run to. Should I stay? I stayed.

There was a terrific bang. I seen a white flash and don't remember any more because of all the bricks and slates which fell off the stables on top of me. That flash was the colour of lightning.

There was a feeling of noise coming down on your head. I felt as if my head was going to come off (I have to wear a hearing aid now), and I felt that noise on my head for days. And vibration. I didn't go unconscious, but bricks cut my hand and there was glass in my hair.

Mr – I've forgotten his name – was getting the horses out of the stables. The first was a big horse. I heard him say: 'Get 'em down, Prince.' But Prince wouldn't put his hooves down near me – maybe he knew I was there.

I had to kneel to get up. Then I felt pain in one knee. I thought I'd lost the other leg, but it must have been numbed. I had been wearing ear-rings and a coat. My hands were dirty. I wiped my hands on my clothes, and found I had nothing to wipe them on. All my clothes blew off with the blast.

Naked, Mrs Owen took herself off to the hospital. No use waiting for an ambulance – no ambulance could get into the area, for all the houses were lying in the roads. She couldn't go home, for she knew there was no home any more. But she knew she was cut and had to get to hospital, so she walked naked through the winter night amid the devastation, over piles of dusty rubble.

She met an ARP warden who said: 'What's up?'

'I've been cut.'

The warden put his coat round her (he collected it from the hospital later).

They stitched me up. I said: 'I got no clothes.' So a nurse went and found me some clothes from a ward – one red woolly jumper miles too small, one navy-blue skirt ten times too big. They didn't keep me in, so I went back. Climbed over brick rubble to a brick shelter. My mum had gone in there, with fourteen-year-old Cyril and my son.

This ancient pram was in there. I put me Mum in that and

dragged her over brick rubble to Arundel Street and then on to my eldest brother's, in Hereford Street. Mum was still hanging on to the bird, although she was crippled and in a pram. We stayed there the rest of the night and then we went to an aunt in Sultan Road, and spent Christmas there and ate the bird.

Some of their friends were not there any more. The Chalmers, for instance, who lived at the dairy in Conway Street. They had been blown up and the front of the dairy was lying in Conway Street.

We moved to Northbrook Street, near Sultan Road, but that house came down on top of us in the landmine blitz of 27 April 1941. 'That's why I think the Conway Street explosion was a plane loaded with bombs, not a landmine,' explained Mrs Owen. A landmine was a cylinder about six feet long based on the shallow water ground mine dropped from the air with a parachute attached to slow its fall and prevent damage to the mechanism; used on land it made only a shallow crater, the main force of the blast being above ground.

George and Margaret Cobb lived at No. 4 Copenhagen Street – three of the roads running into Conway Street from the west were called after Nelson's battles. They had two children, fifteen-year-old Walter and seventeen-year-old Lilian (now Mrs Lilian Prescott). That night the sitting room was gay with Christmas decorations and Mrs Cobb had popped round to their local pub, the 'Hector', on the corner of Conway Street. The family liked to be together in a raid, so when the sirens went, Lilian said to the two men: 'You'd better go round and tell mum the sirens have gone.'

Margaret Cobb came back so they were all complete, sitting round the fire in the gaily decorated room. Lilian was sitting on the left of the fireplace, her father on the right, but nothing seemed to be happening, and she got restless. 'I'm going to wash my hair,' she announced.

Then came a draught, a swish and the house fell on them – all in the fraction of a second.

Everything was black, you could smell gas. The worst part of it was, everything was so quiet; you were just frightened to move. My mother and brother were unhurt, but Dad and I knew we'd been hit. I'd been caught on my left side, he on the right side. Father was the first to speak, saying: 'Are you all right?'

I'll never forget it. I remember hearing people shouting and seeing a torch. An ARP man was asking: 'Are you hurt?' And I replied: 'Yes, I am, so far as I know.'

Then they got us out and took us to the street air raid shelter. Oh, the sights in there. One woman – her eyes simply blown out and lying on her cheeks. The metal entrance to the shelter had simply gone. A lot of people were there, confused, all covered in brick dust. Mr Baines, who was bedridden and lived at No. 2, he got out alive, but Mrs Kimble at No. 1, I don't know what happened to her. Then an ARP man said: 'We'd better go up the hospital.'

We walked, but it was frightening, because the telephone wires were down and there was so much chaos, you didn't know where to put a foot.

The hospital was pretty well packed and we had to wait our turn. We felt shocked and I was terribly frightened. The doctor took us in turn to look at our injuries. There was congealed blood on my head – he could see that (only a small cut but there was also a piece of glass on the back of my ear which was there for a year before it came out. I said: 'It's me arm.' (I had two cuts on my left arm, and still can't use the wrist properly). But the doctor said: 'Do you hurt anywhere else?'

I said: 'I think I've got one on my leg.'

At that, he ripped my trousers right down – just like that!

I'll remember seeing that cut as long as I live. It was awful. So deep that the blood was just pumping out. Must have been something pretty sharp to cause those cuts.

The doctor said to a fireman: 'Take this girl straight upstairs.'

I remember being on a stretcher and being taken to the operating theatre, but nothing more until coming round after the operation, with my mother sitting at the side of my bed. 'Oh, Lily, you have been out a long time.'

'Let's have a mirror,' I said.

I took a look. 'Oh, my God!'

I looked a right sight, with brick dust and congealed blood down the left side of my face.

The leg wound was so deep, they couldn't stitch it, they had to plug it. I cried when I was told the dressing had to come out, I was so afraid of the pain. Father visited me. He looked terrible, like death warmed up. He couldn't eat and was fed on tablets. He was hurt on the right arm and thumb, and was in hospital longer than I was. While we were there, one person who'd been badly burnt in a fire raid was brought in – it was just a charred body, no hope. When the time came to send me back to Portsmouth, I said 'Oh!'

The doctor said: 'I hate to send you.'

We moved to Alpha Road, and used to go out of the City to shelter at Havant every night. Before I got hurt, I never used to worry about air raids; but now I felt numb. Dad heard of a job at Chichester, with a cottage provided, so we moved there. I wanted to go into the ATS, but they wouldn't accept me because of my nerves, so I got work in a factory making exhaust pipes for aeroplanes.

The family never went back to Copenhagen Street, where they had lost not only their home but all their possessions. They did not even know what it looked like until, years after, they saw a photograph of the devastation printed in the Portsmouth *News*.

Mr and Mrs Percy Smith had much the same experience, for they lived in Copenhagen Street too, and their house also collapsed on them. They went back to look next day, and could hardly believe they had got out alive. Mrs Smith had had a presentiment, 'an uncanny feeling that a bomb would be dropped', so she had hidden the children's toys downstairs. But they lost everything material that belonged to them, receiving £85 compensation. All the same, they felt thankful. Mrs Smith still has the scars of her injuries, and her daughter was so shocked that she could not speak for four days; nevertheless they were all alive.

Civil Defence was tightly organised and by now considerably experienced, but the Conway Street area presented new problems: it was virtually impassable for vehicles, difficult and dangerous to penetrate on foot. And it was a very dark night, speckled with the flames of small fires burning here and there in the ruins. Most of the houses had open coal fires, and the explosion had blasted flaming coals across the wrecked rooms. Then there was gas escaping and a gas main burning, and live electric cables exposed in the ruins. Wally Farnes, leading a four-man firecrew, spent the night running out hoses through ruined houses to get at other ruined houses which were burning, or damping down the ground around a blazing gas main. He was wet through, and with a keen wind coming over the dockyard wall, was frozen stiff by morning. He could not even drive the Buick which towed the trailer pump and the vehicle's wheels had to come off to remove the endless lengths of telephone wires which had wrapped themselves round the axles.

George William Batterham was with the ambulance service. Many other ambulances, civilian and naval, had arrived before him, but a very big area had been flattened. 'A shambles', he called it. Many of the houses were only lathe and plaster and the walls one brick thick, he said. Many people were buried under the mountains of debris. In Conway Street itself, he rescued a little boy from a kind of cavity in the rubble, where something was holding up the remains of the roof and preventing them from collapsing on him. The boy could not move because he had been crippled. He had been listening to a radio placed on a table which the blast of the explosion had flung at him, striking his knee and breaking his leg.

Section Officer Donald McGregor of the AFS found plenty for his firemen to do. Two sailors were injured and trapped in the ruins of a public house, and being affected by mains gas escaping from a service supply. The cry of 'Gas' which young Alan Dart had heard, may not have referred only to the vast cloud of brick-dust which had welled up after the explosion; gas really was escaping from a number of points, a hazard to those trapped under wreckage. The firemen cleared enough rubble for a nurse to get through to the injured men and give emergency treatment, and eventually they were stretchered away, semi-conscious, one of them cursing the Jerries to hell and beyond.

McGregor was now joined by Sergeant H. Hicks of the regular Portsmouth Fire Brigade. They were making a reconnaissance when McGregor heard a child whimpering. The wailing was coming from the shaky upper part of a damaged house. Clambering up to the top floor they found a baby about one year old, trapped in its cot by fallen roof rafters. McGregor eased the rafters off the cot and passed the child down to Sergeant Hicks, who tucked it inside his coat for warmth. Then both men descended to the ground by clambering down mounds of rubble, where they were met by a Special Constable who ticked them off for using handlamps. As the lamps were masked in the manner approved by regulations, the fire officers told him where to go. Sergeant Hicks said he was going to take the baby home for his wife to look after, rather than hand it in to a rest centre, but McGregor never heard what happened to the baby, or even if its parents had survived.

Almost at once he was asked by a Warden to do something about a smouldering fire in a house two streets away. The wrecked front door and hall were impassable, so McGregor broke in through a front window. As he jumped down, his feet landed on a large, round, fleshy object which for a shaken moment or two he thought was someone's head. Directing the flashlamp downwards, he saw that it was someone's Christmas dinner bird, which must have been blasted like a cannonball from the kitchen at the back to the window at the front of the house. Similarly, the contents of the fire-grate had been blasted across the hearthrug, setting fire to it and to the floorboards. As McGregor left to get assistance, using the window once more, the whole house erupted into flame behind him. Had he stayed even a few seconds longer, he would have been engulfed.

After making sure that the fire would be confined to that house, and not allowed to spread, McGregor carried on with his tour. He was amazed to find a horse standing trapped amid wreckage with debris up to his withers, but apparently uninjured, in a massive area of flattened buildings and rubble. Probably this was the ruins of the brewery stables, where Mrs Lilian Owen had been injured and stripped naked, and someone was trying to get the horses out.

Later, recalled McGregor: 'I saw a youth who had been out for the evening arrive home to find it a heap of rubble. He went berserk, tearing into the rubble with his bare hands to find his parents. Regretfully, I did not know if they had survived or not.'

In the dark, the full extent of the damage was not clear – it might have been a stick of big bombs. In fact, nineteen streets had been destroyed by a single explosion and 1500 people rendered homeless. Damage had been caused at points up to two miles away from the impact. One of the first people to find the crater was George Rex, a part-time ARP warden responsible for the area designated G1. He had just checked a shelter when he heard a whizzing, whistling sort of sound that went on for say a minute; he thought it might have been a complete 'bomb rack' from a plane. He got down on his face when he heard that, and then stood up and walked to the site of the explosion. 'It was a huge crater. I went down to the bottom of it, the sides were twelve to fifteen feet high and it was thirty to forty feet across. It fell on the corner of Conway

Street at the "Hat Box", a lady's hat shop. My area, G1, didn't cover it. It was in G2, I think, and the wardens there reported it. So I went back to my post on the corner of Church Street.' The strict territorial demarcations of the ARP system worked well enough for a sustained attack but were not appropriate to a single, giant explosion of enormous force.

At the time, Lionel Whitelock, aged seventeen, who lived with his family at the opposite end of Trafalgar Street to Conway Street, was down in the shelter under Semaphore Tower in the dockyard; he felt the hot blast of high-pressure air come down into the shelter. At about 4 o'clock next morning, before dawn but with the pitch blackness of the winter's night getting lighter, those in the shelter were told that there was extensive damage in the Conway Street area and that anyone living near there could go home. 'There was not much apparent damage until you got to Unicorn Gate,' he said:

> Then, as you came up, you could see damage further on. When I was just inside Trafalgar Street, then I could see the damage to our home. The front door was off its hinges and because the windows were blown in there was glass inside the house. But the canary was OK, his cage hung just in front of the window; and the glass cabinet with glassware inside it was untouched, although tons of soot had come down the chimney. I checked that everyone was all right, but I thought then that the damage was superficial. I went upstairs, saw it was dangerous there, and wouldn't let anyone go up.

His father was still on duty in the harbour, so young Lionel was taking command. Undoubtedly, he was suffering from delayed shock. The tension of walking down a street towards your own home, when you know it's been damaged but you don't know yet how bad it is, is indescribable. You keep control, and then you explode. I know, it happened to me once, in 1943.

Lionel's first reaction was, after having checked that his family were alive, to go out and help in the rescue work.

> It was then that the damage struck me. I was shocked. It was devastating. I was sickened at the destruction. Only one building was recognisable – St Agatha's Church at the top of our road (it's still there, the only one left). Next door to it was

> Conway Street School – that was gone. At once I realised that friends must have been killed. Lots of them, because whole streets had gone.

His young sister Edith, who was with him, saw tears streaming down Lionel's face.

He began work in Charlton Street, off Conway Street by the dockyard wall and the torpedo depot. The wall, which is a loopholed fortification, had had its upper ten feet or so pushed back by about a foot.

> Lots of people had cellars – I handed down cocoa to some of them, and some were got out alive next day. I could see little fires burning, but there was such a wide area enclosed in devastation that the firemen couldn't get near it. I could hear people crying and shouting for help. Then someone would say: 'Be quiet.' We'd listen, then someone else would say: 'Someone over here.' And the sights. A hand poking out here, a leg there. When daylight started coming, then you could see the real extent of the damage, and that the ruins were smoking, were very hot. I remember the top of a girl's head – severed. She had long, fair hair. Edith thought it might be a girl she knew, it was near the school. But you couldn't tell the age of anyone, and you didn't recognise anyone, for it was dark and they were covered in dust. When Dad came back from firewatching, he just gathered us all in his arms. 'Thank God you're safe,' he said, and he was crying, too.

At about dawn, Elsie Godding (now Mrs Pitman) and her mother, who ran the 'Unicorn' pub, were released from the Royal Portsmouth Hospital after their cuts and shock had been treated. Their father had suffered only cuts and had remained behind in the bombed pub, from which the injured sailors were being removed. 'At daybreak we made our slow and laborious way home, dreading what we would see,' recalled Elsie Pitman.

> It was, as it happened, worse than anything that could be imagined. Streets of houses and pubs and small shops gone, wiped out as if they had never been. Trafalgar Street, Copenhagen Street, Nile Street, Duncan Street, Conway Street – streets too numerous to mention destroyed, and many friends perished, and this all with *one* explosion. At the bottom of

Flathouse Road, near Pitt Street Swimming Baths, was the biggest crater I've ever seen. It was as deep as a street and as wide as two streets, a big black gaping, smoking hole, surrounded by litter and complete death and devastation.

We got home and found the 'Unicorn' – no windows, no roof, no doors, but the fire blazing merrily, as if nothing had happened, in that big coke stove, and the Christmas decorations still hanging up, but crazily. The bar was full of people, wrapped in Mum's best blankets, and Dad – quite drunk. He thought we were dead! It's a wonder, knowing him, that he didn't say: 'And where do you think you've been all night?'

There was much speculation, as to what had caused the explosion. Some said it was the first landmine to fall on Britain and the authorities were too afraid to tell us. Others say, and this I believe, that a bomber loaded with bombs had been shot down. This makes sense as the guns were banging away at an aircraft shortly before we were all knocked out.

Mr F.G. Hayward was another who did not get home until dawn, because he was on ARP duties in the dockyard that night. It was with fear that he approached the house where he lived with his father, but the 'old chap' was still alive, trying to clean up the house. He and his wire-haired terrier Prince were begrimed with soot and plaster; the stairs to the upper bedrooms were coming away from the wall, and there was no upstairs, not even a ceiling, let alone a roof. 'A cold grey dawn revealed utter destruction; poultry and pet pigeons were fluttering aimlessly about, and here and there remnants of treasured Christmas presents lay about, and tattered decorations wavered gently among the ruins.' Working parties, some of whom were weeping, laboured in the debris and the crater was gigantic, 'big enough to engulf three or four buses'. Mr Hayward thought the damage had been caused by a landmine intended for the dockyard.

I heard about Conway Street at once, because my father was a doctor at the Royal Naval Barracks and part of his province was the Pitt Street swimming baths. The first rumour was that the explosion had been caused by a landmine. The second rumour concerned the finding of pieces of enemy aeroplane near the crater; it was supposed to have been shot down by a night-fighter. I did not visit Conway Street until 21 January 1941, when I described it tersely: 'amidst tremendous devastation, a long oval hole of great depth'. That, I thought, rather argued for a Heinkel 111 complete with full bomb-load; I had been inside a Heinkel and knew the bombs were in fore-and-aft bays which might tend to make a crater longer than it was wide. It was specious reasoning, but nobody was offering anything better.

The Captain of the Dockyard reported to the Admiral Superintendent, Portsmouth, that on 23 December 1940:

> A very large bomb, salvo of bombs, a land mine or crashed aircraft containing bombs exploded outside the Yard at the junction of Copenhagen Street, Conway Street and Charlton Street, flying debris fell on the Offices of the Torpedo Depot, and the blast effect was terrific . . . Production will be at a standstill until the buildings have been made weather-proof, debris removed and night shift will be suspended until the blackout has been repaired.'

He also recorded the dockyard casualties as thirty-four slightly wounded, eight dockyard men seriously wounded, four Servicemen seriously wounded.

The outside civilian casualties in the Conway Street area were given variously. The War Diary of the Civil Defence Warden Service listed fifteen killed, many injured, 1500 homeless, nineteen streets *completely destroyed.* The local Municipal History of the War recorded eighteen killed, fifty-nine seriously injured, 500 admitted to Rest Centres. The London Home Security Intelligence Summary No. 956 for the period 18.00 hours 23 December to 09.00 hours 24 December 1940 informed persons in authority that: 'Some damage occurred in the docks at Portsmouth, but Service casualties were slight. In the city, small residential and business property was affected and over 200 people were rendered homeless. Three people were killed and 44 seriously injured – a few people are still trapped under wreckage.' Some future historian unshakably convinced of the veracity of official documentation may fasten on that figure of 'three people killed' in order to establish that everyone else was guilty of exaggeration. It is an instructive thought. Even more alarming was the discovery of a retired Police Inspector whom I consulted. When he turned up the War History of the Portsmouth City Police Force, he found that it devoted three lines only to Conway Street, referring to it without corroborative detail as a 'very bad incident', but giving pages and pages to the fire blitz of 10 January 1941, full of interminable detail. Having a 'suspicious mind', he said, that looked like a cover-up. It was something to check.

I thought I had found a cover-up, too, when I tried to locate the actual documents on which the local official history had been based. I was informed that they had been destroyed by the author after he had completed the manuscript. 1940 was quite an historic year really, and one would think that basic war records would have been carefully preserved; instead they had been deliberately disposed of. Was there anything which happened in Portsmouth around that time, which might have been mentioned in connection with the raids, that some people might think better burnt? The untold story of the so-called 'Yellow Cavalry' (the nightly exodus from Portsmouth into the country) came immediately to mind, but I dismissed this when I discovered that at national level copies of those records, and many like them, appeared not to have been preserved officially. It was extraordinary, but it was so. Some people apparently had no sense of occasion.

At the grass roots, however, the size of the tragedy was fully indicated. The parish register of St Agatha's Church, the only building left standing in Conway Street, showed that the population of the parish was reduced from approximately 6000 before the explosion to about 3000 after it. Half the population had been evacuated.

What was never established was the cause of the disaster.

In September and October 1965, the Portsmouth *News* printed a selection of letters under the headline 'CONWAY ST MYSTERY BLAST'. Eight of the witnesses opted for a bomber coming down with full bomb-load, two thought parachute mine, two could give no solution. A third alternative was given by one correspondent, Mr H.C. Thompson, who had been the Head Warden in charge of the Conway Street area. He suggested:

> I believe that a German bomber on special assignment to the city, carrying a super parachute block buster, released his load on target. The parachute failed to open, hence the terrifying loud roar caused by the flapping parachute during the bomb's rapid descent. I believe the aircraft got safely away. Not one piece of aircraft, metal, or fabric was found after intensive reconnaissance of the area affected.

In 1972 as part of a feature on Portsmouth in the raids, the *News* printed stories from twenty-three witnesses of what had happened at Conway Street. The majority could offer no solution. Of those who did, five opted for a crashed bomber, one for landmine, and one ingeniously for a bomber which had been hit and jettisoned its load, which must have included a landmine which exploded in mid-air.

In 1979 there was another flurry of correspondence on the subject headlined: 'EXPLOSION MYSTERY'. One of the witnesses, Jack Garrett, had been a customer at the 'Unicorn' pub, and he wrote: 'For years there was much argument as to what had caused it: could it have been a landmine or a plane that had crashed with its load of bombs? If the former, it was obviously intended for the Dockyard; if the latter, a tragic accident. But has there ever been a satisfactory, official explanation?'

Another correspondent, Vernon Mussell, was certain in his own mind. He had been on lookout duty at No. 1 First Aid Depot at St Mary's road bridge, and saw the AA guns open fire at an aircraft coming towards Portsmouth from the east:

> There was a sudden larger glow among those bursts which sped westwards towards me, increasing in size as it came. As it came nearer and lower there was a whine at first which turned into a horrifying scream – the most unearthly noise I have ever heard. I thought it was about to hit our building. The scream died away and, a second or two later, the premises shook as Conway Street and the surrounding roads were devastated. There has never been any doubt in my mind that the explosion was caused by a plane being shot down by the gunners at Hayling. Of course, its cargo may have included landmines or sea-mines, but as far as I can recollect the first landmine came down some time later in Hayling Avenue. Perhaps someone who served in Civil Defence control can confirm this.

No one did.

By this time, it was twenty years since I had last fully accepted the 'downed bomber' story as being the only likely explanation. A landmine it was not. As Mrs Lilian Owen succinctly put it: 'A landmine used to clear the top. This was a crater.' The parachute braked the big cylinder, so that the projectile did not penetrate more than a few feet and all the blast went sideways and up.

Conversely, a really deep crater could only be produced by a V2 slamming in at well over the speed of sound, or a large armour-piercing bomb dropped from a great height, and neither of these caused much blast damage.

But ever since I had researched for my Battle of Britain book in 1959, I had a reasonable alternative. While interviewing a Duisburg policeman, Willibald Klein, who in 1940 had served in the bomber unit KQ26, equipped with Heinkel 111s, he showed me as a curiosity a photograph of a special Heinkel, the Mark 111h, carrying externally a gigantic 2½-ton bomb for use against the Maginot Line forts. He also had photographs of the craters they caused; they were truly gigantic – I had only ever seen two or three of that size. A bomb as big as that, descending at an angle along the track of the aircraft, might well make an oval-shaped crater too.

So I wrote off to the Bundesarchiv at Freiburg in West Germany for information, citing the single Staffel of KG26, which I knew to have been capable of carrying a 2½-ton bomb at that time.

Meanwhile, I had come across another smaller mystery within the greater. Several witnesses, including a police constable, Edward Hair, had vividly described a bomber coming in towards Portsmouth dockyard from the west, from the direction of Stokes Bay and Gosport – and not from the east, from the Hayling Island direction. Constable Hair was actually at Gosport at the time and could describe how he got down in the gutter as the plane flew low overhead.

> Within seconds, there was a terrific roar as the bombs exploded. I went to the side of the Market House, expecting the [Gosport] shipyard to be devastated, but instead saw [across the harbour] the huge columns of smoke, fire and debris rising from what I considered to be the south end of Charlotte Street. It must have been a moonlit night; I could see it all so clearly. It is my firm opinion it was a German bomber which came in over Stokes Bay, and striking Conway Street. I am fairly sure the plane crashed with its bomb-load.

A moonlight night and an approach from the west, instead of a dark night and an approach from the east; and many columns of smoke and debris rising from near Portsmouth dockyard instead of a single column. A good, solid witness – the police constable – on the one hand; and two good witnesses at least on the other, a

Royal Marine sergeant and an ARP man, actually on lookout duty, plus many others who testified to an approach from Hayling Island. How did one explain that?

I do too much research, my wife (and my accountant) always complain. It's not economic. But in this case, it paid off. I didn't just open the Admiralty file at 23 December 1940 while working in the Public Record Office; I thumbed backwards, too. And there it was: document 477, dated 22 December 1940, headed 'Attack on Portsmouth Dockyard'. A single bomber had dropped a stick of bombs into the South Camber, by Semaphore Tower, sinking five cutters, pinnaces and launches, and blowing one Naval rating into the cold wintry water – 'Shocked by temporary immersion' read the Admiralty report. The policeman on the Gosport side of the harbour, cut off from the scene by water, had confused the two attacks by single aircraft; easily done, because the explosions would be almost in line with each other, when seen from Gosport.

I contacted two dozen witnesses for extended accounts and the dockyard men among them inclined, as did Walter Jeakes, to suspect that a special device had been intended to wipe out something in the dockyard, probably the torpedo shop. Even as things were, damage at three major ports had caused a shortage of torpedoes for the submarines at Malta. R.J. Cuss was emphatic that 'the big raids didn't do the damage isolated raiders did.' Basically, they were right. This had been an attack by a single bomber sent over on a special mission. But like the British records, the German records were defective, too, with extensive gaps in them (the great raid on Dresden accounted for many Luftwaffe documents and the Allies soon afterwards captured many more). What I wanted, a detailed report from low down the chain, did not exist; there were only two accounts high up and therefore brief. There was an Air Fleet 3 report:

> *Portsmouth: III./KG 26 durch 1 He III (Sonderauftrag) um 19.49 Uhr mit 1 Max. Detonation beobachtet. Einzelheiten der Wirkung wegen 9/10 Bewölkung nicht auszumachen.*

Translated, this means that III Group of *Kampfgeschwader* 26 using a single Heinkel 111 on special mission reported the detonation at 19.49 hours German time (6.49 pm, British time) of one 'Max', a 2500-kg (2½-ton) thin-walled blast bomb. Results were not observed because of nine-tenths cloud cover. (British reports of the time of detonation at Conway Street varied between 6.45 and 6.48 pm, so the German estimate of 6.49 pm agrees.)

An additional scrap of Luftwaffe information sums up the attack near Portsmouth that night as: '*1 Flugzeug, 1 SC 1000 1 Max*.' That is, one aircraft carrying two bombs. The smallest was a 1000-kg (2400-lb) thin-walled heavy bomb called a 'Hermann' which without fins was six feet three inches long, with a diameter of two feet two inches. The other bomb was a 'Max', two-and-a-half times the weight of a 'Hermann', a real giant of a bomb for 1940, and also '*spreng zylindrisch*' (thin-walled). It was exactly the bomb Willibald Klein had shown me a photograph of, carried by a Heinkel 111h of his own unit, KG 26. And it was certainly the type of bomb that hit Conway Street instead of the Torpedo Shop, 100 yards beyond.

One additional mystery still remains.

C.M. Jeram, who had been with the trailer pump crew inside the dockyard wall between Pitt Street and Unicorn Gate, went off duty the following morning, which was Christmas Eve of course, and midway between his shelter and the torpedo shop he noticed a piece of coloured rope hanging from roof guttering damaged by the blast. He took a piece home:

> It is a length of woven silk cord, which is what the Germans used to connect the parachute to their landmines, and I still have that grisly relic. You will pardon me under the circumstances, but it is a beautiful piece of work, made of silk, green and white, dozens of threads plaited to form a rope. The Germans found that stranded silk gave the elasticity, with strength, to stand the wrench when the 'chute opened.

The only parachute cord I had, for comparison, was from an American paratrooper's 'chute picked up on the battlefield east of Nijmegen in Holland, not as a souvenir but because I could make good use of it. Mr Jeram's relic was in mint condition and immensely thick and strong, quite capable of slowing a mine or a big bomb. Instead of attempting a documentary check, I went to see Cecil Boniface, who also lives on Hayling Island, knowing that he had a piece of parachute rope from Hayling's landmine attack of 17 April 1941, and Mr Jeram having cut me off a piece of the Dockyard rope, I was able to compare the two. They were identical in texture, material, shape and size. But there was no comparison between the comparatively minor damage caused by the landmine at Hayling and the widespread devastation at Conway Street.

H.C. Thompson, head warden there, had suspected 'a super parachute blockbuster', the failure of the parachute to open fully explaining the loud, terrifying roar as the bomb came down. Or did the Germans bind the two bombs together – the Hermann and the Max – to make a $3\frac{1}{2}$-ton combined load? If so, it would explain why although parachute cord was found, there was no report of any parachute material being picked up. And that was a very popular item, silk being scarce.

Today, only St Agatha's Church stands. Conway Street and most of the little streets around it have vanished under a vast dockyard vehicle park. The community has been scattered to the winds.

Chapter 10

DEATH OF A PRIME MINISTER
General Sikorski: 1943

It was not a spectacular crash. On the night of 4 July 1943 the Liberator, all four engines roaring at full power, sped down the runway at Gibraltar and was airborne with 500 yards to spare in spite of her heavy load. As its navigation lights dwindled, two of the VIPs watching thought: 'Well, there's another valuable cargo safely on its way.'

When, instead of climbing, the navigation lights began slowly to descend towards the surface of the sea, they were not perturbed. That trick, to haul a Liberator off the ground and then put the nose down to gain a larger safety margin of speed before beginning the climb proper, they knew to be the trademark of the plane's experienced first pilot, Flight-Lieutenant Edward Prchal, a Czech officer now employed by RAF Transport Command.

But the lights continued to descend steadily and on an even keel. Then all four engines stopped and a split second later there came the noise of impact, as the aircraft hit the sea less than a mile away. The airfield searchlights, doused during take-off to avoid dazzling the pilot, now came on and for a few minutes illuminated the wreck before it sank.

The watching VIPs felt totally helpless, although one of them, General Mason Macfarlane, was the Governor of the Gibraltar fortress and the other, Air Commodore Simpson, was the senior RAF officer there. Because of prevailing sea conditions the high-speed rescue launches had to be moored in the western harbour, five or six minutes away; the only immediate means of rescue was a small dinghy kept on a slipway. There was nothing to do except wait.

When the first launch returned, there were three bodies in it. Two of them were terribly mauled about the head and were dead. They were General Sikorski, Prime Minister of Poland and

Commander-in-Chief of the Polish armed forces serving with the Western Allies, and his chief-of-staff, General Klimecki. The third man was still alive, having suffered comparatively minor injuries. He was the pilot, the Czech, Edward Prchal. Mason Macfarlane, who knew him well and had been chatting to him just before the take-off, noticed one extraordinary fact. Prchal normally never wore his lifejacket, preferring to hang it over the back of his seat. But this time he had worn it, and not hurriedly thrown on but with 'every tape and fastening properly put on and done up'.

The wrecked plane, still spewing oil, was located next morning lying in twenty-five feet of water, and salvage diving began immediately. Among the missing were the Prime Minister's daughter, Madame Lesniowska, a British liaison officer, Colonel Victor Cazalet, a Polish courier from German-occupied Warsaw, and two British secret service men. The local shallow-water diving team, using improvised breathing gear in imitation of the Italian 'frogmen', was commanded by William Bailey, twenty-one years old, and included among its members Lionel Crabb, twenty-seven, who was to become a famous underwater operator and died in 1956 while diving under a Russian cruiser in Portsmouth harbour. First, they tried to recover bodies, then they received a direct order from Mason Macfarlane to search for a portfolio of secret papers, and finally they were plagued by an accident investigator who wanted them to note the position of the controls in the cockpit.

The Liberator no longer resembled an aircraft but a collection of dangerously jagged bits of metal. Scattered on the sand amid the wreckage were suitcases, zip-cases, boxes of Turkish Delight (a favourite of Sikorski's daughter), medals, a cigarette case, a jewel-box, and a crate of Leica cameras. Bailey and Crabb saw what they at first thought was a headless body hanging from a wing, but it turned out to be an empty overcoat. It was Crabb who found Victor Cazalet, still strapped to his seat. In visibility of about twenty feet, out of the corner of his eye, he saw the man nodding at him, hair moving with the water, one big open eye staring.

Three of the dead were never found – the General's daughter, the co-pilot, and one secret service man. Possibly they got out, or were helped out but drifted away in the night, injured, and drowned. Most of the aircraft wreckage was recovered and brought on land for the accident inquiry. The pilot testified that after take-off he had put the nose down to gain speed and then tried to pull back the stick to begin the climb – but nothing happened, the controls would not move, and the aircraft had flown straight into the sea. When he had seen that a crash was inevitable, Prchal had cut all four engines. The wreckage, however, showed no sign of jammed controls and efforts to jam the control wires which run through the fuselage by shifting baggage and other cargo failed.

Sikorski, a fine-looking man of sixty-two, had had a premonition about the flight back to England from Cairo via Gibraltar, and had specifically asked for Prchal as pilot because of his great confidence in him. His daughter had shared his fears of flying and had said that she was going to die by drowning and feed the fishes of the sea. The General's premonition has been passed off as the result of total physical exhaustion after his prolonged tour of Polish units in the Middle East at a time of crisis for Poland.

Premonition, or superstition, also played a part in the return of the bodies to England in the Polish destroyer *Orkan*. Because of the immense heat experienced in Gibraltar in summer, Mason Macfarlane had recommended that the remains be flown home at once, but the Polish government felt that this was not fitting, so they lay in state in the Cathedral for some days. The funeral ceremony, before embarkation, was planned for 8 July. On the evening of the 7th a Polish officer paying a last visit to the Cathedral found the Polish soldiers who should have formed the honour guard inside, standing on the cathedral steps, muttering about *ghosts*. And indeed the cathedral was full of strange noises and a sickly sweet stench pervaded it. Lifting the drapes on Sikorski's coffin, the officer saw that the gases of decomposition had expanded and burst open the zinc-lined coffin. The eerie creaking noises coming from the coffin holding General Klimecki indicated that the same process was well under way there.

In the eight hours remaining before the funeral ceremony the Poles procured new coffins, transferred the remains to them, welded them shut and cleaned and fumigated the cathedral. Afterwards, because of some seaman's superstition, the captain of the *Orkan* was reluctant to take even Sikorski's coffin aboard, let alone Klimecki's, and he resolutely refused to accept the body of Colonel Cazalet, the British liaison officer, who had therefore to be buried in Gibraltar. The captain maintained that to turn a ship into a hearse meant doom for the vessel, but he was overruled, and the coffins of the two Polish generals were taken aboard the destroyer on 8 July 1943. In wartime, losses of destroyers were heavy and it was not altogether surprising that the *Orkan* was torpedoed and sunk exactly three months after, on 8 October.

Less easy to dismiss now were recollections of anonymous telephone calls which three senior Polish officials in London had received some six weeks before, on 26 May. The calls were identical in import, the hearer being informed that 'General Sikorski's plane has crashed at Gibraltar and all its passengers have been killed.' That was exactly what had happened, for the sole survivor was the Czech pilot. At the time they seemed not so much nonsensical as non-informed, because General Sikorski was then on his way out to the Middle East and had landed safely at Gibraltar the previus day. Now they took on a sinister aspect, not as bad history but as an accurate prediction of an event still six weeks in the future when the messages were received.

Of course, a mysterious accident involving an Allied commander-in-chief must alwaya arouse speculation. General de Gaulle, in a similar position and an equally if not more controversial figure, now no longer travelled by air when in Britain because of an incident which might well have been sabotage, and Sikorski himself had nearly been killed in Canada when both engines of his aircraft cut out on take-off. Even if both cases had turned out to be simple malfunctions, propaganda use could be made of them by interested parties. Nothing could ever be straightforward in these matters and as intelligence affairs traditionally remain forever secret, they certainly were not likely to turn up in the official histories of the guilty party. And if the guilty party was holding the accident inquiry, with equal certainty nothing was going to come to light there, either. Therefore it is necessary to be completely objective in seeking out motive. Opportunity is another matter.

In that sphere the British had the best opportunities, as Gibraltar was a British base and the aircraft belonged to the RAF. The options open to the Soviet Union are not clear, as the only known NKVD agent at Gibraltar was Kim Philby. The Axis powers had a fact-finding and sabotage unit installed on Spanish territory opposite Gibraltar which so far as is known concentrated largely on the mining of ships by underwater operators, usually Italians. An aircraft was a more difficult but not impossible target.

In the matter of motive the possibilities are fascinating, because so very many people had reasons of varying strength and urgency for desiring the death of the Polish commander-in-chief and head of state in exile, while as many again could hope for advantage by accusing others of sabotage and assassination, even if it had been only an accident. For the Liberator had dived into the sea at a time of crisis among the Allied powers in regard to Poland.

The tragedy of Poland is that she lies between two mightier powers which are often at loggerheads, Russia and Germany. And at this time it was Stalin's Russia and Hitler's Germany with which any Polish leader had to deal – a choice so terrible that it was hardly a choice at all. The Poles had only recently blotted their copybook with their Soviet neighbour by inflicting on the triumphant Red Army its first humiliating defeat when in the 1920s they had resisted a Russian takeover. Then in the 1930s, backed by Britain and France, they had resisted Hitler's claim to Danzig (now Gdansk). The result had been disaster, for Hitler the supreme anti-communist had made a pact with Stalin to divide the wretched Polish nation between Germany and Russia. This had unleashed the Second World War, brought France into the dust and Britain to impotence momentarily. Then Hitler, as he had always wanted to do, had launched his armies eastwards across Poland into Russia. And it was what they found there, in particular at the Forest of Katyn, which provided deadly ammunition for Germany's propaganda offensive designed to split the uneasy alliance between Stalin's Soviet Union and the Western powers, on which the fate of the world now depended.

At Katyn near Smolensk the German invaders of Russia uncovered the mass grave of some 4500 Polish officers murdered in about 1940 as part of the Soviet Union's normal policy of eliminating the bourgeoisie so as to better control the workers themselves. It was not a war crime because the Soviet Union was not at war with any nation at the time of the massacre. Many thousands more Polish prisoners-of-war from 1939 were, and still are, missing, presumably buried in mass graves which have never been found. The Germans invaded in 1941, but by 1943 the tide of their advance had turned and was ebbing. They desperately needed to divide the nations arrayed against them, and as their military superiority waned, so diplomacy and propaganda became more important. The announcement of the chilling discovery was made by Berlin radio on 13 April 1943, and was central to the Sikorski affair, as was the advance of the Red Army back towards Poland.

In 1939, the Poles who had escaped from the invasion and defeat of their country by the combined forces of Hitler's Germany and Stalin's Russia were fleeing from them both. When the Russians entered the war in 1941 (because the Germans attacked them) thereafter the Poles had enemies on both sides of the lines. When in 1943 the Germans announced the news of Katyn, the Russians looked likely to swamp Poland shortly, to enter the homes the Polish volunteers had left behind them, with results which the Poles could grimly predict. Of course, the Poles knew well that the Germans were equally capable of killings such as those at Katyn, but this realisation would not make the Russian atrocities go away. The effect on the morale of the large Polish forces fighting with the Western Allies, mostly in the Middle East, was serious. This was one of the main reasons why General Sikorski flew out to the Middle East in May 1943 to visit his armed forces. By now it was also becoming clear that the Soviet Union would not recognise the Polish government in exile, but intended to enthrone their own communist puppet regime as soon as their undisciplined forces had conquered the country from the Germans.

This situation had further repercussions in that the British and American governments were allied to Stalin as well as to Sikorski, and Sikorski had many divisions fewer. It was not unlikely that they would sacrifice him in order to keep the Russian army in motion on their side. By July 1943 Sikorski had toured the Polish units in the Middle East and solved certain disputes with General Anders, the local Polish commander. He was flying home from that mission when he died in the inexplicable accident at Gibraltar. Russia had just broken off diplomatic relations with him and his government.

Within twenty-four hours Dr Goebbels had his radio stations broadcasting to all Europe the 'news' of the Polish premier's assassination at Gibraltar by the British Secret Service. The allegation was cleverly linked with the actual assassination in 1942 of the French Admiral Darlan in North Africa (although the French themselves, encouraged by General de Gaulle, seem to have organised this murder). However, the German message came out loud and clear: national leaders of small nations fighting for the Allies, beware the knife in the back!

Interestingly, Stalin himself fomented the same idea later on. He confided to some of the Yugoslav communist leaders that he had been told by the Czech leader, Benes, that the British 'set Sikorski in a plane and then neatly downed the plane – no proof, no witnesses!' This tale, reported by Milovan Djilas, a Yugoslav communist who later fell out with Tito, has rather the tinge of what is called 'disinformation'.

More immediate and perhaps believable was the reaction of the travelling Russian diplomat Maisky. By some comical error of the British Foreign Office a Russian diplomatic party led by Maisky was routed eastbound through Gibraltar at the very same time as Sikorsky's Polish diplomatic party was passing through westbound, and in the period immediately following the official breaking-off of relationships between the conquering Russians and the Polish government-in-exile. Mason Macfarlane, the Governor of Gibraltar, who was required to offer hospitality simultaneously to both parties, prevented an impasse by being frank with the Poles, asking them to stay in bed late until he had hurried the Russians on their way early. When Maisky heard of this later and realised that Sikorski had been killed at Gibraltar shortly after his party had left, he was said to have concluded that the British had hurried him away the better to kill Sikorski quickly.

No doubt he had been reading Niccolò Machiavelli's *The Prince* in bed. From long experience of government the Florentine had come to the conclusion that 'all men of modest life, lovers of justice, enemies to cruelty, humane, and benigant' are likely to come to 'a sad end' if they achieve supreme power. Only the men and women of opposite temperament survive. Consequently, any successful wielder of power would be perfectly capable of ordering the death of Sikorski for sufficient reason of state. If he was not, he would not be in power.

Diplomatically, the death of Sikorski served both British and Russian purposes: it was the effective end of an independent Polish government-in-exile; the rump that was left could not embarrass either Churchill or Stalin. One fancies however that the Russian need to eliminate Sikorski was much stronger and more urgent than any British feelings on the matter. From the propaganda viewpoint, the crash served Dr Goebbels and the Nazis excellently well in sowing mutual suspicion among uneasy wartime allies, just as Stalin's words would have engendered similar suspicions among the powers he was technically allied to, but would soon confront.

But Sikorski might also have been the victim of a Polish faction which, enraged by Katyn, regarded anyone still prepared to talk to the Soviets as a traitor to be eliminated. Nor must one disregard the fact that few governments are genuinely monolothic; often there are divided opinions as to the best courses to pursue and in what direction the national interest really lies; further, in politics, diplomacy and statecraft men rarely say what they mean but more frequently some devious alternative.

While the speculation went on and the propagandists made use of the tragedy, inquiries were taking place at Gibraltar. Those responsible for protecting the aerodrome had the most to fear from any searching investigation, because it is well nigh impossible to guard an airfield of any size with less than 10,000 men. But it is possible to guard particular aircraft, and Sikorski's Liberator, it turned out, had been well looked after, with a guard outside and inside. Only authorised personnel had been allowed to approach it or to go inside. This virtually ruled out sabotage by either the Germans or the Italians.

The RAF Inquiry could find no evidence of the controls having been blocked, either accidentally or deliberately, as the pilot claimed. But other statements made by Prchal which could be checked were confirmed by other witnesses or by the wreckage itself. Nor had there been anything difficult or peculiar about the weather conditions. The suggestion, made first by the Germans as an immediate insinuation, that the Czech pilot had been employed by the British to deliberately crash the converted bomber was contradicted by the evidence brought to the surface. Instead of a comparatively gentle slow 'ditching' of the Liberator, the wreckage showed all the signs of a crash at fairly high speed, just as Prchal described, which in spite of his survival must appear a somewhat suicidal proceeding. Most conclusive of all, it turned out that Prchal had not been chosen by the RAF to pilot Sikorski; it was the General himself who had asked for Prchal specially because he had been impressed by his flying on a previous occasion. Nor, in spite of the recoveries of cargo from the seabed, had the Liberator been overloaded; the weight had been within the permissible limits.

That Prchal had been found floating with the aid of an inflated lifejacket while everyone else was dead, drowning or dying of dreadful injuries appears to be a case of freak survival not infrequent in wartime and not unknown in peace. That Prchal never normally wore his lifejacket but left it draped over the back of his seat disturbed Mason Macfarlane, who suggested some sort of blackout or momentary aberration as a possible cause of the crash. Prchal himself must have been concussed and very likely could not recall everything which had occurred. The Inquiry exonerated him, the sole survivor, and there matters must stand – as an RAF informant told me in 1980, 'a real mystery'.

As with an active volcano the matter slumbered for many years until in the late 1960s there came a series of eruptions which dissipated a great deal of heat while shedding a mainly lurid light. In Winston Churchill's obituary broadcast to Poland on the death of Sikorski he had said: 'Soldiers must die, but by their death they nourish the nation which gave them birth.' It was not one of his happier phrases, and a German author called Rolf Hochhuth fastened on it as the title for a play, *Soldiers*, in which he accused Churchill of having had Sikorski assassinated as a kind of prelude to committing mass murder in the German cities by his bombing policy. As the proponents of the British bombing policy were mostly still alive, as were those of their victims who had survived the firestorms, this was a doubly incendiary theme for a playwright at that time.

But it was directly in the Hochhuth tradition, for in an earlier play he had accused the Pope of knowing about the extermination of the Jews in Hitler's camps while the massacres were still going on, and yet failing to protest for fear of antagonising Hitler, who then was the real ruler of Europe. This was brave of Hochhuth, for there were more Roman Catholics in the world (many of them in Poland) than there were members of the British Cabinet and Air Staff. Nevertheless his plays did pose moral questions which touched on the experiences and dilemmas of many who had gone through the Second World War, and who knew that it is often far from easy to do the right thing, even if one knows exactly what that is.

Of the British production of *Soldiers*, Harold Hobson was to write in 1980 in the pages of the *Sunday Times*, that it was 'a play that contained a thoughtful discussion of the ethics of area-bombing'. 'Thoughtful' was not a word one would apply to the discussions which the play provoked on television, notably in the

David Frost shows, over a period of a year or two. The TV impressario gathered together some formidable casts which included the Czech pilot of the Liberator, Prchal; Randolph Churchill, the son of Winston Churchill; Kenneth Tynan, the London impresario who wanted to produce *Soldiers* with Laurence Olivier; David Irving, a British author who wrote a 231-page book on the Sikorski crash titled *Accident*, in which he hinted that it was no such thing; Carlos Thompson, an Argentinian author who wrote a 461-page book called *The Assassination of Winston Churchill* designed to prove that Hochhuth was not an historian; and in one programme Hochhuth himself off-stage on the telephone.

Even allowing for showbiz razzmatazz the single programme I happened to see was quite extraordinary in the fury and confusion it generated; it seemed to me that someone was protesting far too much. One supposes that the reason most probably was that the matter was not at that time old history but instead live politics. Even so, one was left with an unquiet feeling. If there was no fire somewhere down there in the smoke, why so much heat?

Chapter 11

ABANDONED
Flights without a Pilot: 1915–1945

The story of the abandoned ship which survived its crew is symbolised by the tale of the *Marie Celeste* because of the exceptional element of mystery. Most instances of this sort merely go to prove that your best lifeboat is the ship; don't leave it unless it leaves you. Even so, it's chilling when in real life you come across an empty boat drifting in the seaways; even if you take it in tow and find out eventually that it merely broke loose when empty, and there's no story.

One might think that such things could not occur in the air, that an empty aircraft or one with the pilot dead would crash and be destroyed at once. Very often, this is so; but not always. Sometimes there are freak flights which are almost unbelievable.

Oswald Boelcke described what is still the classic case in a letter home from the front in September 1916. Leading a flight of five Albatros scouts over Bapaume he had spotted and stalked a similar formation of British Martinsyde single-seaters. As soon as he had cut off their retreat he attacked, shooting down a Martinsyde in the first minute or so. It just went down 'like a sack', he said. Then he chose a second victim and a turning fight began, in which Boelcke, who was deadly with his guns, thought he had settled his opponent's hash several times over; but still the Martinsyde went on circling. It simply would not fall, although he was hitting it repeatedly. Baffled, Boelcke crept in close until from a distance of a few feet he could look in to the cockpit. There was the explanation! The man was hunched over, probably dead, jamming the controls so that the pilotless aircraft was kept in an endless turn. He made a note of the Martinsyde's number, 7495 – which eventually identified the pilot as Lieutenant S. Dendrino – and left the dead man circling. On returning to base he found that another member of his flight, Sergeant Reimann, had also claimed Martinsyde No

7495; to avoid any possible ill-feeling, Boelcke suggested that neither of them should claim the ghostly biplane as a kill.

In those days before parachutes it was rare for an airman to leave his machine in mid-air, although it was possible to fall out accidentally, leaving an abandoned aircraft to fly itself. Quite the most extraordinary case occurred to L.A. Strange, an RFC pilot, in May 1915. He was flying a Martinsyde scout also; the only good thing about them, he thought, was the Lewis gun mounted on the top plane which allowed one to fire forward over the airscrew. At 8500 feet over Menin in Belgium he was fighting an obstinate Aviatik two-seater whose observer was potting at him quite dangerously with a pistol. One drum from the Lewis gun having failed to quell his opponent's ardour, Strange reached up to change drums; but it was jammed. He then held the stick between his knees so that he could use both hands on the drum but it would not shift. Then he raised himself in his seat to get a firmer grip on the drum, and a number of undesirable things happened. The stick slipped from between his knees, his safety belt slipped down, and the Martinsyde, which was staggering upwards on the point of a stall, flicked over on its back and began to spin.

Strange slipped right out of the cockpit and was hanging by both hands to the drum of the Lewis gun which, with his aeroplane, was now above him. Belgium lay 8500 feet of thin air beneath his boots, and all that attached him to the plane was the drum which he had just been trying to take off the Lewis. Now Strange prayed that it would stay jammed, firmly.

In spite of the spin, Strange managed to transfer first one hand from the drum to a centre-section strut, and then the other; and then with incredible, kicking acrobatics, managed to get both feet back inside the cockpit (which was above him) and boot the stick so that the Martinsyde came upright. Then he fell back into the cockpit and found himself sitting so low that he could not see out over the sides. The stick wouldn't move, either. The reason was that his fall back into the cockpit had smashed the seat and he was now sitting on the floor, jamming the control cables. Strange throttled back and wedging himself upward, wrenched loose and threw overboard all the smashed bits of seat which were helping to foul the controls. By now, he had used up most of his original 8500 feet and pulled out over German-held Belgium under fire and with little height to spare.

On both sides of the lines that day there were mysteries to be cleared up. The German observer, a conscientious and reliable man who always submitted totally objective reports, claimed a victory over a British Martinsyde which spun down north of Menin after the pilot had been thrown out of it, although he was not actually seen to fall clear. The Germans spent half a day searching a wood near Menin for the wreckage of the Martinsyde, without avail. Rude remarks were made about the conscientiously accurate observer.

For his part, Strange brought back an aircraft with all the instruments on the dashboard kicked to pieces, the seat cushion missing and the seat smashed up. It was not the result of enemy action, but only a world-class acrobat could possibly inflict such damage on his own machine, and even then it was difficult to see how! After all, to kick the dashboard in, one would have to be outside the aircraft – and who was going to believe that? The only evidence was the way Strange shambled around for days afterwards with an aching back.

Only something similarly unbelievable could explain the mystery of the L-8. She was a small training airship, some 250 feet long, based at Treasure Island Naval Air Station, San Francisco. At 6 o'clock on the morning of 12 August 1942, she took off with a two-man crew for a routine flight over the Pacific. The pilot was Lieutenant Ernest De Witte Cody, an experienced airshipman aged twenty-seven. His co-pilot, Ensign Charles E. Adams, was making his first flight as an officer, having put in twenty years as an enlisted man. They were anything but novices, and the chances of enemy action on the West Coast of America were almost nil. But at 07.50 Lieutenant Cody radioed the control tower to report sighting a large oil slick five miles off the Farallon Islands. He was going down to check.

The crews of two fishing boats which were nearby later testified that when they saw the airship nose down to 300 feet and begin to circle the oil slick, they expected depthcharges. So they hauled in their nets and got clear. But the 'blimp' dropped nothing. Instead it rose upwards until it entered cloud, and vanished. At 08.50, fifteen minutes after Cody had reported the sighting, Treasure Island tower could no longer raise the L-8.

Two hours later, she was sighted drifting in over the coast by Fort Funston. Some anglers fishing in the surf tried to catch her dangling draglines, but could not get close enough. However, they could see that the cabin door was open and the gondola apparently empty. The blimp drifted off inland, dropped a depthcharge on the Olympic golf course (which failed to explode), and, sagging in the middle, approached Daly City, a southern suburb of San Francisco. A passerby paused to photograph the unusual sight, for the helium gas inside her had contracted, so that the weight of the control car was distorting her belly; L-8 looked like a heavily-pregnant sow. Normally, a blimp maintains its shape in such conditions, because the crew operate a scoop which lets air into a compartment called a ballonet, thus making up for the contraction of the helium in the gas-filled bag.

At about 11 o'clock that morning consternation struck the centre of Daly City. An airship was apparently trying to land in the main street! Windows banged open, heads popped out, Traffic came to a standstill. Fire engines, sirens wailing, sped through the streets. Police officers rehearsed what they were going to say to the 'crazy Navy barnstormers'. Unbelievably 250 feet of collapsing airship came down in the street, missing the two-storey buildings, and settled neatly on top of two cars.

Police and firemen approached the control car and looked inside through the open door. As far as they could tell, it was all absolutely normal. Everything seemed in place, and even the radio was switched on, as if the crew might be back in a minute. But there was no sign of the crew!

'My God!' somebody exclaimed. 'There's nobody here! The damned thing landed by itself!'

That was too much to swallow. If the men weren't in the control car, they must be in the gasbag up above it. That made sense and the firemen, using their axes, slashed open the gasbag – which promptly deflated. 250 feet of airship swiftly shrank to an up-ended gondola and the control surfaces where the tail had been. And still there was no sign of a crew.

A naval salvage party arrived within the hour. Technical inspection on the spot drew a total blank. There was nothing whatever wrong with the L-8 (apart from what the firemen had done to it), and nothing whatever was missing – apart from the two-man crew and their lifejackets. Even the liferaft was still in position, untouched, and that was the first thing one would launch if the ship had to be abandoned over the Pacific. But there was no reason to abandon – the engines were OK, there was plenty of fuel, the feed lines were open, the ignition-switches were on; and the throttles were open, one fully open, the other at half-speed – which indicated a turning manoeuvre. The radio was on – although radio communication had apparently been broken off by 08.15. Most significant of all, the confidential portfolio still lay in the control car. This held orders and other secret or confidential documents which it was a court-martial offence to lose! There was no trace of fire, nothing whatever to indicate that an emergency had occurred.

The local press got some good copy and excellent quotes. 'There was something eerie about it – we got chills down our spines and we couldn't wait till we got out of there,' a fireman was said to have told a reporter.

'What really gave me the creeps, though,' alleged another man, 'was this sandwich with one bite out of it and a mug of coffee that had spilled on papers on the desk. The papers were like a blotter and the coffee was still warm!'

An intensive search was organised off the Farallon Islands, but in spite of the good visibility and the smooth seas which had replaced the early morning overcast, nothing was found. The fishermen, who had been watching apprehensively for the first depthcharge to fall from the blimp onto the oil slick, declared that they saw nothing fall out of the L-8. No depthcharge. No body – or bodies. The blimp had simply gone up into cloud and vanished. Cody and Adams were never seen again.

Convincing explanations are lacking, as with the *Marie Celeste*. It has been argued that because the men had been wearing their lifejackets, as regulations laid down, then their bodies at least should have been found, buoyed up by the jackets. But a human head in an endless expanse of waves is a very difficult object to search for; a negative report does not mean that nothing was there, merely that nothing was seen.

Of course, there were suggestions that the two men were no longer on planet Earth, having been removed by some extraterrestrial force, but this is rather like using a steamhammer to crack a couple of nuts. The most probable explanation is to suggest something inherently improbable – like Strange falling out of his aeroplane and then climbing back in again – but without the happy ending: for instance, that one man could perhaps have leaned out, the other holding him – and then some small, silly thing might go wrong, like a sneeze at the critical time, or an unexpected movement by the ship. Why the men should want to lean out is not so easy to understand – but the door of the control car *was* open when L-8 drifted across the surf line and the fishermen plunged to grab the draglines.

Empty airships flying themselves are one thing, empty aeroplanes quite another, especially when they are bombers traversing enemy territory under fire from the ground guns, to which a big aircraft flying straight without evasion must be an ideal target. But it has happened more than once. Two of the most bizarre cases both involved the products of Dr Dornier, the twin-tailed bomber known as the 'flying pencil' because of its low-drag fuselage. The first incident occurred in 1940.

On 22 October the *East Anglian Daily Times* issued a warning to its readers. The crew of a Nazi bomber were loose in the area. Members of the public were asked to keep a sharp lookout and report anyone suspicious to the Chief Constable of East Suffolk. The Home Guard had already spread the same message. The plane had made a belly-landing on mud flats at Erwarton in the early hours of 21 October, after passing through the gun defences of the naval port of Harwich, which had brought it under heavy fire. But there was no sign of the crew.

Some 130 miles and five counties away, at Shaftesbury in Dorset, the authorities were faced with an equally puzzling problem. They had a German aircrew without an aeroplane.

The Germans also were amazed. Having baled out over what they believed to be France, hopefully near their base in Brittany – but which was in fact Salisbury Plain – they had trudged off into the night, only to find at the first village that all the shop names and notices were written in English, not French.

Their target had been the port of Liverpool, midway up the west coast of England; their Dornier Do 17Z had belly-landed on the east coast of England near Ipswich; and they had baled out over Salisbury Plain in the south of England. And they had, during the night, once actually returned to their base near Cherbourg without knowing it!

A Suffolk schoolboy, Christopher Elliott, who was a keen aviation enthusiast and aircraft spotter, was given a piece of this Dornier by a girl cousin after it had been taken to Ipswich. (In 1944 he refused the offer of a blackened souvenir from a Liberator, not knowing then that it came from the mysterious disaster which had killed Joe Kennedy.) After the war, one of the local, Suffolk incidents he researched into was the mystery of the Dornier which had landed itself on the mudflats, empty.

The story he obtained was this. From Brittany, the pilot planned to approach Liverpool by flying over the north coast of Wales – a sound idea, as the coastline would give a good 'fix'. Then he would turn eastwards along the Welsh coast and pick up Liverpool. But he must have turned east too early and met a severe magnetic storm somewhere in the centre of England which put his radio out of order; he then turned south, ran into a second storm, and this time had his compass affected, without his knowledge. He flew south to a position which must have been near Cherbourg, not far from his base, when he dediced that his erratic compass was indicating that he was headed north. So he turned 180 degrees and flew back to England! Over Salisbury Plain the Dornier met a third magnetic storm. The compass went haywire and he had no radio with which to ask for bearings, but at least, so the pilot thought, he had brought the machine back to France, where they could bale out. He put the Dornier on automatic pilot and then followed his crew out of the escape hatch, while the bomber flew on over England, a target eventually for the guns of five counties.

Their lack of success in 1940 was not surprising. The number of guns was small and their shooting was bad even in daylight; at night they had nothing to aim at unless the searchlights could fasten on a bomber or condensation trails in bright moonlight picked out the paths of the enemy. In 1943 and 1944 the situation was very different. Instead of burning a city steadily all night, one bomber droning in every minute, and perhaps one shot down out of hundreds overhead, the typical German night raid became a scalded cat affair – in and out fast and high, and someone shot down almost at once. A parachute held in the searchlight beams for ten minutes or so; or a flying glow that became a roaring furnace and then finally fell to fiery pieces, as if dripping to the ground. And no one getting out.

It must have been intimidating to the German aircrews, who were required by Hitler and Goering to fly until they died or were smashed up, to face such hazards in pursuit of mere 'retribution', on operations lacking all military value and which could not succeed. We called it the 'Baby Blitz'. I saw it mainly from London, where I was stationed, or from my home near Portsmouth on weekend leave. In both places rocket batteries were in spectacular action, as well as radar-equipped night-fighters. It was clearly very dangerous to be a German airman.

This point was realised particularly clearly by the all-NCO crew of a Dornier 217M which took off from a KG 2 airfield near Paris on the night of 23 February 1944 with a dud starboard engine which was losing ten per cent of its power, not enough to warrant aborting but certain to reduce performance significantly. Instead of crossing the coast at 18,000 feet, they could only make 15,000 feet. At that height, they were quickly 'coned' by the searchlights and so proceeded on towards London in the most public manner possible through a sky swarming with cannon-armed night-fighters, as if they were cavorting across a floodlit stage. The pilot, Herman Stemann, sheered off a bit to the west, where the sky looked blacker.

The final approach to their target was to be made in a dive to 13,000 feet from 18,000 so as to streak across central London, where the multitude of flaming rocket tubes waited in Hyde Park inside the ring of high-velocity 3.7 AA guns which encircled the capital, and with the night-fighters stalking outside the circle. But this Dornier could only reach 15,000 feet to start with and was suicidally down to 9000 feet as it approached the heart of the gun area. They turned away to the northwest rocking in the blast of the explosions. The lights on the instrument panel went out; there were streaks of reddish flame spurting from the labouring starboard engine. The pilot and observer held a quick conference. Both expressed the same opinion – this machine would never get home at this height through those defences. Stemann put the Dornier on automatic and the four men baled out over Wembley in Middlesex.

Nearly 300 aircraft were over that night, and not all went to London. Over Portsmouth area, where I was, they were either very high or very low; some on their way to London, others trying to get out of the beams and under the spectacular AA fire by diving. One plane came down with bellowing motors and the sound of the wind screaming past him; then the pilot cut his engines and did a violent turn to the left and his engines roared again, harshly and determinedly. To the north, someone baled out. In a hopeless attempt to baffle the radar they were dropping strips of black paper about two feet long, one of which I picked out of the gutter. One could fairly call the defences fully alerted.

Even in Cambridge, fifty miles north of London, they had an air raid warning 'Red' and some people went to shelter. Mrs Jane Riglesford was one of them. When the 'all clear' went and she came out of the shelter, she found a bomber at the bottom of her garden. A Dornier 217M. An army officer who was staying in the house strapped on his pistol and went out to arrest the crew, but they were already in police custody fifty miles away. The Dornier was quite empty – except for its cargo. The complete bomb load, two tons of incendiaries in three containers. Some of the fire bombs were of the new and cunning type containing a small amount of explosive, which I soon had good cause to remember well. The fuel tanks still held 600 gallons. The bomber had made a soft wheels-up landing (unheard by people sheltering in the garden) after flying through fully alerted air defences for fifty miles and unable, because abandoned, to deliberately jink or alter course or height. The perfect target, one would think. As far as is known, this was one of the only two aircraft of any nation during the war to make a good landing when flying itself. Both were Dorniers, although the 1944 version was twice as powerful as the machine of 1940.

Empty aircraft flying themselves were now not all that uncommon, although rarely so docile. I saw two myself shortly after, and appropriately both were American Liberators. I noted the daylight sky changes in my diary: 'In 1940 one saw over one's south coast home mainly the black cross of Germany, in 1942 only the RAF cocardes, and now in 1944, mainly the white star of the USA.' On 11 May, in preparation for D-Day, there was a lot of daylight activity over the coast; Fortresses going out about 1 pm, Liberators returning about 6 pm. I counted thirty-one in the Liberator formation, the lead twenty-eight keeping fairly well together, the rest straggled out as though damaged. I noted that the big twin rudders shone white in the sun, so they may have been Navy. They were followed by a lone Liberator, quite low at 2000 feet, flying slowly inland. It disappeared behind the trees towards Chichester, and there was a rending crash. The crew had already baled out over Bognor, so it was empty when it ploughed into a laundry, killing a fourteen-year-old girl and injuring twenty others. Sensibly, the crew had got out of the damaged machine as soon as they crossed the coast.

The next one was an opposite case, the crew abandoning rather than face the journey back over the Channel. My viewpoint had changed, too, because I now had the Channel behind me. On 10 August 1944, at 1.30 pm, one of the great aerial displays of history took place. We've all seen little biplanes doing aerobatics, standing on their ear and falling off it, and so on, to the angry-wasp sound of a tinny four-cylinder engine. But imagine a great four-motor bomber doing it. The enormous beast would dive, engines blaring, pull up violently in a climbing turn that made a great arc across the Normandy sky, then fall off it in a stall turn; and then repeat the performance; and then do it again. It virtually stopped the war for a quarter of an hour. I imagine that machine had an audience of at least a million: everyone was watching, from ENSA to the Waffen-SS. I only saw one parachute dropping and drifting slowly; somebody who had popped his head out earlier had counted four.

Then the fighters turned up, Ours, commissioned to end the nuisance. They went tearing round each other near the parachute, snapping off three-second bursts of cannon and machinegun. I wrote in my diary, 'The Liberator pulled up in a climbing turn, fell off, but did not spin. Then it stood on its tail at an angle of 70 degrees, climbed several hundred feet, then fell away slowly, spinning to the left, with white smoke trailing from one engine, and went down out of sight below the apple trees that masked the horizon.'

Actually, these manned aircraft now seemed strange to us. Before coming to Normandy I had been in London and seen the first six weeks of a weapon of the future, the robot flying-bombs called V1s – little jet aeroplanes which carried one ton of explosive and no pilot. In the next three weeks the Army I was with was to capture all the launch sites of the V1s. A month after that and I was to be on the arrival point of an even more futuristic weapon – the V2 rocket; before the war ended I was to see the contrails going up vertically as the Germans fired them in our view. We had seen the future, and it worked – if not very accurately then.

During the war some very strange aircraft landed in Britain. On 5 May 1941 a member of the Royal Observer Corps, Mr R. Baker, reported that a twin-engine, twin-boom monoplane with German markings had landed near his post in Suffolk. No such animal, said HQ. But there was. It was a Fokker G-1 fighter, flown by a Dutch test pilot with the director of the Royal Netherlands Aircraft Factory as passenger. They had taken off on the pretext of making a last test flight of the machine before delivering it to Germany, then slipped their German fighter escort by dodging into cloud. In July the same year three Belgians got away in a Stampe biplane, a light aeroplane with a range of only 250 miles. There were even German or Austrian deserters who flew to Britain. One of them was believed to be carrying important information regarding the German jet-fighter programme. He was an anti-Nazi Austrian stationed at a fighter training unit near Innsbruck who got away in a two-seater training version of the Me 109 with long-range tanks in May 1944. After 600 miles he ran out of fuel over East Anglia, made a crash landing, but survived it although injured.

An aerodrome specially designed to take every type of aircraft but particularly those in trouble was Woodbridge, east of London, seven miles from the coast. It had runways two miles long and half a mile wide. Cripples returning from raids with dead pilots or dud engines or half an undercarriage made for Woodbridge. It had everything, including the expensive 'Fido' system for dispersing fog. In the first half of July, 1944 it took a burning British Mosquito which incinerated its crew; an American B-17 Fortress on two engines which managed to land short and deforest a small wood on the perimeter; an American P51 fighter with only one wheel which flipped over and decapitated the pilot; and on the 13th a fully operational, finely equipped version of the very latest in night-fighters – a German Junkers 88G-6B. The pilot, Unteroffizier Hans Mackle, was looking for the big Luftwaffe fighter base at Venlo in Holland, and thought he had found it. However, as his petrol was almost finished, he was willing to land almost anywhere.

Quarantined away at one corner of the enormous airfield was a top-secret American unit with seven B-17 Fortresses, very old and weary army planes loaded with a quarter of a million pounds of explosive, destined for a project called Aphrodite, which had succeeded in turn other top-secret projects called Azon and Double-Azon. The American crews and ground crews were touchy about the amount of high explosive with which they were cohabiting and they distrusted the British on principle. If the journalist who told their story is to be believed, they thought they were the only people on that base capable of recognising a Ju 88 by sound. Actually, all except the tone-deaf and deprived could tell British from German because of the unsynchronised engines among other things, and of course the locals had heard more Ju 88s than those Yanks had had hot dinners (as the saying went). There was a distinctive wrrrrrm-wrrrrrrm to the German motors which once heard, was never forgotten.

According to their journalist the American pilots disappeared under beds and cupboards in a flash, knocking over the card table and chairs as they did so. All except for one realist, the card player with the winning streak, who was dubious of the degree of protection afforded by a bed or a bit of plywood.

While the prone people waited for the first bombs to come howling down on them, there was a hissing noise and a great brightness outside. There were cries of anguish.

'The Fido lights! The Limeys have turned on the Fido lights.'

'Jesus Christ,' exclaimed another voice. 'Can't they tell that's a German plane?'

'They're showing the pilot where to bomb!' raged another.

The menacing drone of the Ju 88 as it did a careful circuit of the now invitingly-lighted aerodrome mixed with American blasphemies regarding dumb limey bastards. Then the sound of the German motors above them changed to the snaps and pops of their being throttled back. An American parted the blackout curtains and saw a big twin-motor plane, wheels down, flashing past. British sentries stopped the Americans from seeing what happened next. What they had done was to help the German land and let him taxi in just as if he was on a large continental airbase. When Hans Mackle had taxied to a position from which he could not take off again, a member of the ground staff, Sergeant Kenneth E. Clifton, made the copybook capture. The easy thing to do, which the Americans wanted done, and the British themselves had done on an earlier, similar occasion, was to let fly with bursts of trigger-happy firing likely to scare off the visitor. The hard thing to do was to hold your fire and capture, not a wrecked aircraft, but an intact one, as had been done in this case.

And this case was very important. The Junkers was fitted with the latest type of Lichtenstein SN-2 radar which could not be jammed by the metal-foil and paper strips known as 'window'. With an intact instrument to examine, the British had the answer in ten days – 'long window'. The machine was also fitted with *Naxos* and *Flensburg* sensors. The array of aerials put the wind up the Americans, according to their biogropher. A major declared: 'That son of a bitch has got *everything*, and there's no telling what he radioed back to Germany. Our security is shot, and we've got to pull up and get out!' And get out they did.

Part of the Americans' anger had been because they had been told they were at Woodbridge to save Britain from Hitler's new robot weapons across the Channel by carrying out an equally weird scientific counter stroke which was chancy in the extreme. 'They told us that we had the most important mission in the European Theater, the number one priority, and by order of General Eisenhower himself.' The British appeared not to give a damn about it. The Americans were too young and naive to understand that they were in a power game of US Army and Navy politics, which was to involve a President's son and a President-to-be in a tragedy so incredible that it had to be hushed up for many years.

Chapter 12

INTO THIN AIR

Joe Kennedy: 1944

The Kennedy's were always competitive. The boys not merely had to play games, they had to win. They were to be competitive among themselves, too, not merely against others. Joe, the eldest, was to be President. He was vigorous and athletic. His younger brother Jack had problems with his health. He was to be the literary one of the family, upholding its cultural banner. Bob was to be a lawyer and Teddy was too young. Thus Joseph Kennedy Senior and his wife Rose had decided. Their ambition was a very great one, although they already possessed two of the essentials – extreme wealth and a passionate concern for politics. But there were two drawbacks to overcome: they were Irish and Catholics. They would be sallying forth from an Irish enclave to storm the bastions of Anglo-Saxon Protestant America and capture its central citadel of power.

Everything a Kennedy did was politics. Either it helped him towards the White House or it did not. Joe Kennedy Junior visited Moscow under Stalin, Milan under Mussolini, Madrid besieged by Franco, Berlin under Hitler as the German armies marched into Poland, and was in London in time for the British declaration of war in September 1939. There his father was US Ambassador and the Kennedys lived like Royalty.

Joe Senior wanted to keep America out of involvement in Europe's battles; Roosevelt wanted in. Eventually, after the bombing of Pearl Harbor on 7 December 1941, two years and three months after it started, Roosevelt got America into the Second World War. The two men's views represented large bodies of opposed electoral opinion. By backing the losing side, Joe Kennedy Senior had dropped a few points politically. It was for his sons to take up the running. They had to succeed in the war, against other men and against each other, and with more to lose, because they were millionaires too. Ironically, for two and a half bitter years, Joe Kennedy Junior, handsome, confident, arrogant, charming, trailed after his younger brother, John F. Kennedy. Both joined the Navy, but in the week that Joe became a Cadet, Jack Kennedy was promoted Ensign. When Joe got promotion, Jack got higher promotion, to Lieutenant. When Joe was sent to uneventful Puerto Rico Jack went off to the Pacific War. While Joe was completing pilot training on Liberators, Jack was becoming a hero, having had his PT boat cut in two by a Jap destroyer, saved his crew, and been missing for a week. All this was going to count with the voters after the war.

In October 1943, at last Joe Junior got his chance. His squadron of Navy Liberator patrol planes was sent overseas to England and employed on anti-submarine patrols in the Bay of Biscay. All Joe had to do to get ahead of John F. Kennedy and his Pacific War medal was to sink a U-boat, but the Germans would not cooperate. By the spring of 1944, Joe had completed a tour of thirty-five missions: he had made four submarine contacts, without scoring a success; had shot down no enemy planes; had in fact seen no real action at all, but had repeatedly risked his life by bad-weather flying over the sea and the hills of Cornwall and Devon.

Because the Germans could use them, many ordinary radio navigation aids were not available in England; the lack of them killed Amy Johnson. Even with their own systems, as we have seen, German pilots had mistaken parts of England for parts of France or Holland. A winter's flying in these conditions, with little to show for it, was a severe trial; but try explaining that to a political meeting. There was a plan to protect the Normandy invasion beaches from torpedo attack by corking up the U-boats; Joe Junior was out flying on D-Day looking for those submarines, but even that got him no action, let alone a medal to match John F.'s. The only consolation was that he was now promoted full Lieutenant, so that at last he was equal in rank to his younger brother in the Service.

To be out on D-Day Joe had signed for a second tour of operations, but he had another reason besides politics. He had fallen in love and wanted to stay in England. The girl had a title – but then most of his British friends had titles; alas, she also had a husband – but he was fighting in Italy. Joe's sister, Kathleen Kennedy, had married the Marquis of Hartington. She worked in

London, her husband was serving in Hampshire in an infantry battalion due to go to Normandy, and Joe was down in the Devonshire mud where his squadron was based. They all met, including the girl, at a kind of midway point between the Hampshire town of Alton and London, a country cottage which they called 'Crash-Bang'. With a nickname like that, one would guess it was near Bagshot. The girl it seems had Yorkshire connections, and sometimes stayed on an estate at Sledmere, amid villages with some unlikely names, Fimber and Wetwang. I can swear these names are not fiction, for I served myself in a Scottish battalion near Alton in 1942 and in another at Sledmere in 1943, and I can recall what the general feeling was. People didn't ask how you were, but: 'Are you for the Wall?'

Hitler's Atlantic Wall had been built up by German organisation and Allied propaganda into a nightmare of enormous, concrete defences resembling a science-fiction fantasy. That we had matching devices of massive floating concrete I knew, for I had seen them from Hayling Island beach when on leave. I was glad, for I thought we might need them. Everyone understood that a basic question was to be bloodily settled: could the Allies get ashore and make good their lodgement; or would we be driven out of Europe once again?

That was the background to the long-threatened V-weapon bombardment of the invasion marshalling areas in London and on the South Coast, which became actual, fantastic fact in mid-June 1944. There were three different secret weapons involved and their concrete emplacement seemed impervious to ordinary air attack. In 1943 and 1944 the Allied air forces deluged them with 100,000 tons of bombs for the loss of 450 Allied aircraft and 2900 aircrew. There were reasons for believing that the results of the bombing might be negligible.

Destroying cities is easy – you open them up with blast bombs and then fire the ruins with incendiaries. You burn them. The old-fashioned method with new-fangled means. But concrete resists both blast and fire; and thick concrete resists direct hits by large bombs.

Three different, rival and equally futuristic methods were devised to cope with this targeting problem. One was British and RAF, two were American – the US Army Air Force versus the US Navy Air Force. To Joe's Navy squadron at Dunkeswell in

Devon came a request for volunteers to carry out a top-secret mission; a number of men came forward but Joe Kennedy was chosen.

The project carrying the most official 'clout' was undoubtedly that of the US Army Air Force. The idea was supposed to have originated in the brain of a US second lieutenant, been taken up by General Carl Spaatz commanding the 'Mighty Eighth' US Army Air Force based in East Anglia, then been taken over by General Hap Arnold in Washington, and impressed on General James H. Doolittle, who replaced Spaatz as Eighth USAF commander during this period. Back in the USA, masses of men and machinery were mobilised to build replicas of the German secret weapons sites so that they could be bombed and the best methods identified. Resources were unlimited; it was get up and go. In England a mighty force of forty war-weary B-17 Flying Fortresses were documentarily 'lost' so that the real planes could vanish and then turn up at a secret airfield, together with strangely-equipped B-24 Liberators, and a force of 1000 US personnel aided by another 1000 British, Czechs, Poles and French. This was the force which first assembled at Woodbridge before panic at the capture of that radar-equipped Ju 88 led to it being hurriedly moved, bag and baggage, to a new, cordoned-off airfield at Winfarthing-Fersfield in Norfolk, within the East-Anglian empire of the 'Mighty Eighth'.

The reason for its assembly was that someone had had a top-secret idea to pack 'war-weary' bombers full of high explosive and then fly them by remote control into the mouths of the V-weapon sites in the Pas de Calais opposite Dover. Ten to twelve tons of HE all in one mighty wallop! It was foolproof, it was simple, it couldn't fail – but it did, every time, usually killing or maiming USAAF pilots, occasionally descending on England rather than France, rather noisily. The pilots, if they could still talk, were under orders to tell lies to the British about what was going on. More top-secret security.

One notable dud of an officer, got rid of by his squadron, was discovered posing as an irreplaceable expert, busily empire-building in this new set-up, where no one knew anything much so no one was likely to find him out. The situation was ideal for such people to flourish.

Against the sheer size of this magnificently showy edifice, the contribution of Navy planes and personnel, added towards the end of July, made an unimpressive display. Twenty officers, sixteen enlisted men, one Liberator four-motor patrol bomber, two twin-engined Venturas. One of the officers, the one chosen to fly the sacrifice ship, the Liberator, was Joseph Kennedy Junior.

The reason the navy were there at all was not merely the failure of the Army Air Force but its belated discovery, officially by General Doolittle, that far from being a daring experimental application of twentieth-century technology to warfare, the idea of radio-controlling empty aircraft was not new: the navy had been flying pilotless target planes – 'drones' – for ten years. So, even at the risk of putting ten-second experts out of a job, the Navy had been called to the rescue.

Because of the high-level tragedy that was to ensue, the many-faceted background was subsequently suppressed or played down. Firstly, the fact that radio control of vehicles had been perfected long, long ago. Even as a schoolboy I knew this, had actually seen the most spectacular radio-controlled vehicle of all – HMS *Centurion.* She was a 23,000-ton battleship 600 feet long, built in 1911–1913, and after the Great War converted to remote-controlled target ship. Her captain, three quarters of a mile away in the destroyer *Shikari*, could make the completely empty battleship alter course to avoid the salvoes of shells or patterns of bombs, make her slow down or speed up to her maximum of 17 knots, or wreathe herself in a smokescreen coming from her own funnels, simply by dialling code numbers on an instrument not unlike a telephone. This system not merely worked, but still worked even if the target ship was being hit by the bombs or shells, because the control system included eighteen tiny aerials, not just one. The first public and detailed announcement of this technical triumph, which it was indeed in those days, appeared in *The Times* of 9 October 1928. This revealed that the ship, although empty of men, was not quite uninhabited, for the ship's cat had decided to stay aboard during firings and had been joined by three other cats.

Later, the system was applied to aircraft, usually DH Tiger Moths – which could be either landplanes or floatplanes – to give practice to anti-aircraft batteries. They were called drones and were capable of both taking off and landing without a pilot.

But it was the Germans who made the first operational application of the idea, in a commonsense and deadly manner. They made a glider bomb which was really a small, fast aeroplane with a wingspan of about ten feet, radio-controlled from one fast twin-motor bomber, a Dornier 217. They first went into action on 25 August 1943, against a British convoy escort group off Spain, attacking again with better results on 27 August. In the Mediterranean, they ran up an impressive score of warships sunk or damaged, British, American and Italian, including the Italian battleship *Roma*, sunk.

The Americans got hold of an inferior idea afterwards, and on 23 May 1944 used it for the first time. Azon was a bomb which could be guided only to one side or other of the target; no corrections could be made for range – that had to be accurate from the start. Even so, it knocked out bridges when the skies were clear and flak was absent or light. If flak was heavy or visibility hazy or cloudy, Azon was no good.

No good either was the US Navy's first operational use of the German idea. They loaded up some excessively war-weary, extremely slow old torpedo bombers and sent them pilotless by radio control against the Japanese fortress-island of Truk, stiff with flak. Seven pilotless planes were dispatched and seven were duly shot down short of the target. This had happened quite recently, General Doolittle learned, but there was so much pressure for action now if not before, that no one seems to have taken time out to sit down and think. The reasoning was: the targets will take a lot of HE to demolish, so stuffing a fast fighter with explosives won't do; it will have to be a four-motor bomber – the largest and slowest target possible. But with the highest in Washington pushing for results this way, what could those on the spot do but obey?

At least one could say for the Navy planes that they looked good. No worn-out old hacks here. The Liberator, which could carry a greater load than the Fortress, was only five months off the assembly line at San Diego. The two mother-ships, aircraft which came over from the States also, were twin-motor Venturas already equipped for radio-controlling, and had crews experienced in the work. It was only the drone itself – the Liberator – which required conversion, because this was a 'one-off job'. There were three tasks involved: gutting the fuselage and removing all excess weights so as to load twelve tons of Torpex, a new explosive twice as powerful as TNT; fitting a device which would, by radio command, arm the mass of stable explosive and make it 'live', so sensitive that it would detonate on impact; and fitting a receiver to take the flight control commands of the mother-ship and relay them to all the manual controls in the pilot's cockpit, after the pilot had left.

The latter were much more advanced than the Azon-based system devised by Army Air Corps, which General Doolittle had judged to be a 'project put together with baling wire, chicken guts, and ignorance'. Joe Kennedy was given a tour over one of the two Ventura mother-ships, which would be handling him on the day. The drone-controller or 'peter-pilot' faced a radar scope and a television screen which would show the picture being received by a TV camera in the nose of Joe's Liberator. On the control-console was a miniature joystick rather larger than a pen which, when Joe had handed over, would enable the peter-pilot to fly the Liberator. This job required experience and skill, for the miniature stick, the 'peter', could not convey the feel of the Liberator's controls, nor the seat-of-the-pants sensation. Next to the peter-stick were two telephone dials similar to those used years ago by *Shikari* to control *Centurion.* Only on these you did not dial speed or course instructions, or 'make smoke', but the aircraft ritual of 'flaps down', 'cowl flaps open', carburettor heat on'. Below were two bright-red switches, marked 'DESTRUCT' and 'ARM'. At the moment, they were secured by wire, so they could not be knocked accidentally.

This was a critical difference between the method used by the Air Corps (Operation Double Azon, overall code name *Aphrodite*) and the Navy part of the project (Operation *Anvil*). Radio control did not include take-off: there had to be a human pilot. At a safe height he would hand over control of his aircraft to the mothership; and then bale out. The Army used one pilot, the Navy two, but that was no great matter. The key difference lay in the method of arming the nine tons of explosive in the Air Corps Fortress and the twelve tons in the Navy Liberator. The Air Corps used the simple way: just before he baled out, the pilot would manually arm the load, so it was ready to blow on any impact. The Navy used what seemed the safer way, but was more complicated; the pilots baled out, and only after they had gone and dropped safely clear would the controller in the mothership arm the twelve-ton load by sending a radio message. The mother-ship and back-up mother-ship would then follow the abandoned drone until they sighted the target and then, using the TV camera in the nose, guide the big four-motor plane directly into the mouth of the bunker. To discourage German fighters, in place of the machineguns removed to save weight for explosives, broom-handles painted black had been wired into the turrets.

The commander of the Fersfield base, Lieutenant-Colonel Roy Forrest, was a Texan determined to work with all the rival teams, especially the Navy team headed by Commander James Smith, but also with an Army half-colonel, Dale Anderson, in charge of yet another group, this time drone-control experts from Wright field who were to replace the Air Corps' 'chicken-guts and baling' wire method with something slicker. The Air Corps were infuriated because they had heard Navy men claiming that the Navy had been called in to bail out the Air Corps (which of course it had), and the Navy men were incensed because the Air Corps were jibing at their lack of combat experience (most of them had just come from the States and had not been in a combat zone before). The Navy were hoping to get Joe Kennedy away first in Liberator T-11, before the Air Corps could coax the next lot of tired Fortresses off the ground. But it was not to be.

While Kennedy had been up practising handing over control to, and being flown by, the mother-ship, the Air Corps had been readying four new drones. And one of the Navy's crack experts, Lieutenant 'Bud' Willy, had shown how little he knew about combat area flying by taking a Ventura right over the Portsmouth gun area without knowing either the code of the day or the warning signals for a balloon barrage ahead; he got shot at by the one and nearly took his wings off on the other. But by 4 August the Army Air Corps had fitted four B-17s with a new control system amazingly similar to the Navy's – it even included two men and TV – and had them warming up ready to attack four V-weapon sites in the pas de Calais: Siracourt, Watten, Wizernes and Mimoyecques.

The drone destined for Siracourt never cleared the English coast. The engineer baled out near Woodbridge as planned, the pilot was heard to shout 'Taxisoldier! Taxisoldier!' (the code meaning he was handing over control) frantically, but stayed with the aircraft, probably to fix an electrical fault. He never got out. The B-17 spun into a Suffolk wood and blew up.

From the drone sent to Watten both men baled out and lived, one by a mere fluke. The pitch control was not working, and the B-

17 flew through a British balloon barrage and was sent on its way with a hail of British AA fire, to meet a hail of German flak shortly after. It would not dive or climb and there was no 'destruct' capability; if left to itself the drone might wipe out a French town or village. The controller deliberately sent it past a German flak battery, but they clean missed it on the first pass, so he tried another and this time they got it. After the bang, the battery was silent.

The drone targeted on Wizernes nearly made it. Both men baled out, one with a strangled shout of 'Taxisoldier!' His static line failed, when the reserve chute proved defective and he was still manually pulling the silk out of its case when he landed in a tree – which saved his life. The B-17 had run into a flock of B-24 Liberators heading out on a mission, then cloud obscured the target at the vital moment and the drone overshot by 700 feet. The Germans were puzzled by the size of the crater and the absence of any traces of crew or guns.

The fourth drone, aimed at Mimoyecques, flew erratically after the crew had baled out. The pilot was very badly hurt in his fall. The explosion was 500 yards short of the target causing smoke and dust to rise 10,000 feet into the air, and the mother-ship came back to England in the middle of a flock of V-1 flying bombs being busily shot at by British AA which did not stop for the American plane.

The results of the explosions were not nearly so great as had been expected (there had been talk of damage reaching seven miles from impact point). The American report on the drone which hit Watling Wood in Suffolk noted that it flattened the trees for 120 yards around. The main crater was five feet deep and twenth-five to thirty-five feet wide, with a shallower extension about three feet deep to fifty feet further out. The Luftwaffe report on the drone which overshot Wizernes noted that the aircraft had been blown into tiny scattered fragments to distances of 500–1500 metres, that the main crater was twenty metres across and there were other craters of various diameters.

Four drones had gone up, noted Colonel Forrest, and four had come down; but the four targets remained intact, and had hardly been endangered. The fiasco had involved a large force – four B-17 drones, two B-24 mother-ships, two B-17 mother-ships, two navigational B-17s, one radio-relay B-17, a Mosquito radio relay aircraft, USAAF and RAF fighter cover, many small spotter aircraft to see where the parachutists landed, four Mosquitoes, four P-38 photo planes and two generals, Doolittle and Partridge, scooting about in their own fighters to see the fun.

Forrest concocted the sort of bland report that the management like to read, which blossomed up through channels to the head of the US Army Air Corps, General H.H. Arnold, who wired back: 'PRELIMINARY REPORT ON YOUR AUG. FOUR EFFORT VERY ENCOURAGING . . . WE ARE FOLLOWING PROJECT WITH GREAT INTEREST . . .' The General of all the Generals was understood to be as happy as a sandboy with his new project, which he saw as a preview of the future.

On 6 August the next effort was launched. Two drones only were to be used, both aimed at Watten. One B-17 carried nine tons of the new Torpex explosive, copied from the Germans and lent by the British; the other B-17 held 180 big firebombs and 830 gallons of jellied gasoline. Crack open the rocket base with the first one, flood it with flaming napalm with the second. Everything went perfectly with the first drone until its first turning point over the North Sea, when instead of turning right for Watten, it turned left, rolled on its back, and fell into the water (at 51.50N-02.37E). The second drone, trailing ten minutes behind, turned and headed for London, lost itself in the haze, was discovered circling Ipswich, and after much coaxing seaward dived into the North Sea (at 51.32N-02.26E) in a ball of yellow flame.

These futile attacks which killed or maimed pilots over England without hurting the Germans were taking place from the same air base that the new Navy team was using; the news could not be kept from them. Colonel Forrest had a high opinion of the US Navy, he did not expect to find anything wrong with their set-up, but he thought that he would see. Commander Smith showed him over the PB4Y Liberator T-11 which Joe Kennedy and 'Bud' Willy were to fly. Neither of these two commanding officers was technically at all expert in this exotic field, but it was clear to Forrest that the radio controlled equipment was really sophisticated and well-tried in comparison to the Army Air Corps' Azon-based system. But he baulked at the arming panel, which was a plywood lashup of switches and terminals. He told Commander Smith that it looked like something you'd make with 'a number two Erector set and Lincoln Logs'. The Commander replied that it had been made at the Naval Aircraft Factory in Philadelphia, and presumably they knew what they were doing. Of course, it wasn't so well thought out as the rest of the equipment, but that was because they'd had to improvise a new arming panel to do an unheard-of job – blow ten tons of Torpex on impact – he explained. Forrest replied that, for all their experience, that arming system looked 'about as safe as a basketful of rattlesnakes'.

Clever though it was to arm the deadly load by radio signals from a safe distance after the crew had baled out, there was a danger of the plane's receiver picking up a similar signal from another source and obeying that in error. On test in America, it had done just that three times. Here in England the air was full of calls from ground units, aircraft, tanks and ships, with radar, with beams, with jamming. The only electronics expert in the navy unit, Lieutenant Earl Olsen, had discovered that the receiver, as delivered from the States, would pick up signals over an alarmingly wide band; and he carefully narrowed this to reduce the hazard to a minimum. The executive officer, Lieutenant Bud Willy, had had a safety pin inserted into the panel to prevent such an occurrence, but Olsen was suspicious of it, as a mechanical solution to an electrical problem. He argued with Willy and the ordnance officer, but they would not concede his points. The panel had been designed by experts, and that was that. Olsen had the knowledge, but not the rank; Colonel Forrest had the rank but not the electronic qualifications.

Joe Kennedy himself was tentatively approached, but declined to interfere in matters which, strictly speaking, were no concern of his. He had no status in the unit, because he did not belong to it as he had to the patrol bomber squadron. At Fersfield, he was just the 'hired help'. That was his attitude and he would not change it. He appeared admirably in control of himself almost to the last, when unexpectedly the weather cleared and it looked as if the mission could go next day, 12 August 1944. That evening, he was irritable and jumpy, snarling to himself in the hut.

It wasn't nerves, it was something most servicemen have had to put up with often. His girl was leaving the cottage and going up to Yorkshire for the weekend, to the estate of Lady Virginia Sykes at Sledmere; and he had hoped to join her there, the mission behind him. But the bad weather had delayed it and now he wasn't even allowed out of camp to tell her the news, because Bud Willy said he couldn't. The noisy-voiced little ex-chief petty officer jumped up to wartime temporary lieutenant had told Joe Kennedy: 'All you have to do is fall off your bike and sprain your ankle, and the mission's off.' Commander Smith had backed Bud Willy. Lieutenant Kennedy could go down the road to the nearest public telephone box, and that was all. All the Kennedy clan had tempers and Joe was well-equipped in that respect. Fuming, he made his calls. But the only number he could get was his sister Kathleen's, and he knew she was away in America, her husband having gone to Normandy with the Guards. Someone in their house would have to pass the message on. So the change in their arrangements, and the goodbye, quite possibly his last, had to be made via an intermediary.

Most of Kennedy's friends guessed about the girl, knew it was serious, and expected him to marry her on the well-earned leave that would follow the successful Navy mission. But Kennedy was close-lipped about her – none of them knew she was married already, to an Army officer serving now in Italy.

The 12th of August showed promise of being a warm day, something the Americans believed almost never happened during an English summer. The mist rolled away early; the mission was on. There was a final briefing and Kennedy was told the target: Mimoyecques. Inland of Cap Griz Nez, it was difficult to find; its purpose was not known for certain. There was much talk of rockets, even of rockets capable of hitting New York. But the official Air Force staff description of the targets at the time lists them as 'rocket gun installations in the Pas de Calais area'. In fact, Siracourt and Lottinghem were V1 storage and launch sites; Watten and Wizernes were V2 sites; Mimoyecqyes was V3, an incredible weapon which could come under the heading of 'rocket gun', except that there were fifty 6-inch gun-barrels each 400 feet long set in concrete deep underground. Hopefully, the Kennedy drone would impact on the site at 7 o'clock that evening, so that it would be coming out of the sun towards the defending German gunners.

Roy Forrest had trouble making the weather reconnaissance people understand that, for a low-level attack, he wanted low-level weather, 'down where the prairie dogs walk'. Rather than argue further, he took off in his private P-38 Lightning fighter with Commander Smith crouched in the nose, went to Mimoyecques, saw that the weather was good even for prairie dogs, and was back at Fersfield by 5 pm. After that, it could all roll.

The mother-ships rolled first. The lead Ventura, code *Zootsuit Red*, was airborne at 5.55 and climbing to 2200 feet, followed a minute later by the back-up Ventura, *Zootsuit Pink*. A minute later the navigation aircraft to lead the whole armada, a B-17 Fortress codenamed *Zootsuit A-Able*, was away. Then came Kennedy's white-painted Liberator, codenamed *Zootsuit Black*, loaded till the tyres were near flat, slowly rolling onwards almost forever until finally, reluctantly, there was air under its wheels and it was climbing to 2000 feet. Inside the bomber were 374 boxes of Torpex weighing 21,170 lbs plus six demolition charges of TNT of 100 lb each. Much of the load was in the bomb-bays, but some was stacked on the command deck and in the nose-wheel compartment. Kennedy and Willy were surrounded by explosives.

As he climbed slowly to his first turning point at Framlingham, a veritable armada began to form up behind him. There was a Mosquito codenamed *Zootsuit C-Charlie* which was to radio back a weather report from over the target; another Mosquito flown by Elliott Roosevelt, son of Franklin D. Roosevelt, the President of the United States, which was to photograph Joseph Kennedy Junior as he jumped from *Zootsuit Black*; a B-17 Fortress codenamed *Zootsuit B-Baker* which was to act as relay ship in mid-Channel; a P-38 Lightning for high-altitude photography over the target; another P-38 Lightning with Colonel Forrest the base commander and his passenger, Commander Smith. Then there was a P-51 Mustang fighter flown by General Doolittie, commander of the Eighth Air Force; five more Mustangs as close escort, and more still higher up; and there were probably a few unlisted rubberneckers, too, who saw this amazing armada and nipped over for a quick look. One thing was almost, but not quite certain – the British AA defences were unlikely on this occasion at least to mistake American planes for German.

The plan was not to hurry the mother-ships. Control by radio signals would be carried out during the flight from Framlingham to Beccles to check out all control channels except that which would 'arm' the explosive cargo. When Beccles was reached, the mother would order the robot to reverse course towards Clacton, turn it out over the Thames Estuary where Amy Johnson had died, and bring it down to Manston aerodrome in Kent, where Kennedy

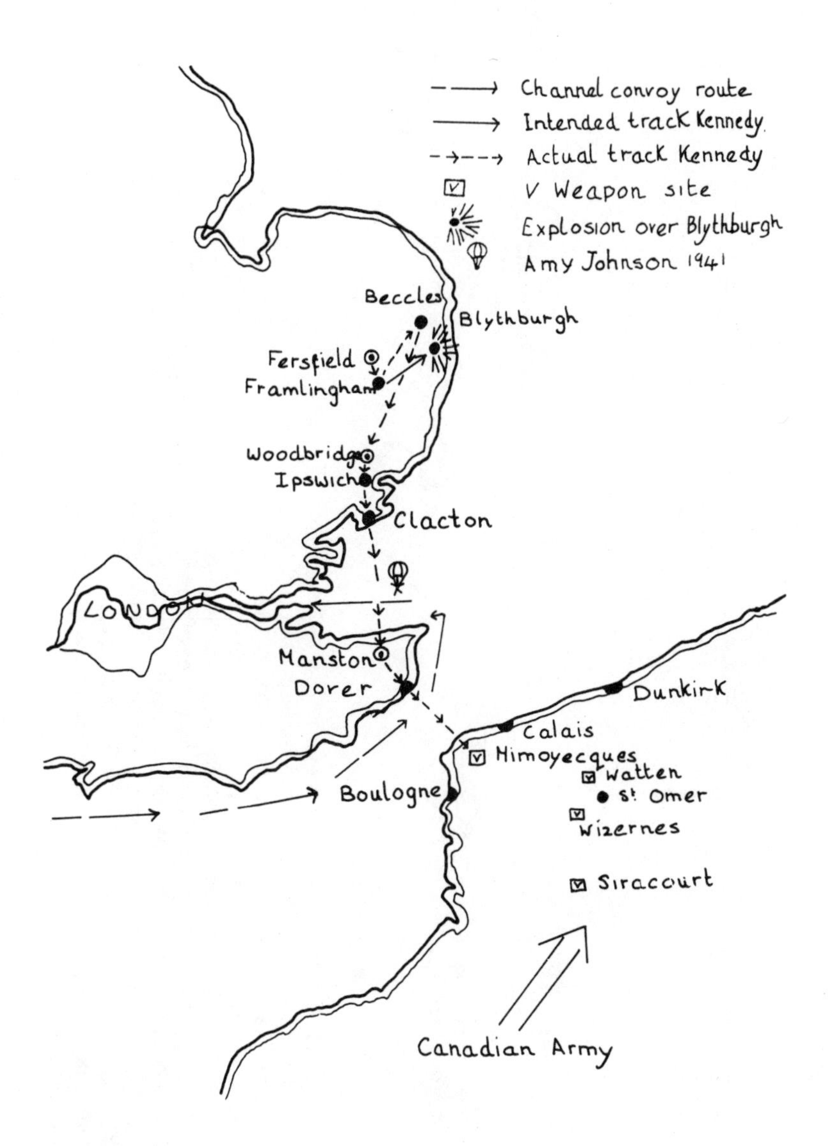
Channel convoy route
Intended track Kennedy
Actual track Kennedy
V Weapon site
Explosion over Blythburgh
Amy Johnson 1941
Beccles
Blythburgh
Fersfield
Framlingham
Woodbridge
Ipswich
Clacton
LONDON
Manston
Dorer
Dunkirk
Calais
Himoyecques
Watten
St. Omer
Boulogne
Wizernes
Siracourt
Canadian Army

Kennedy

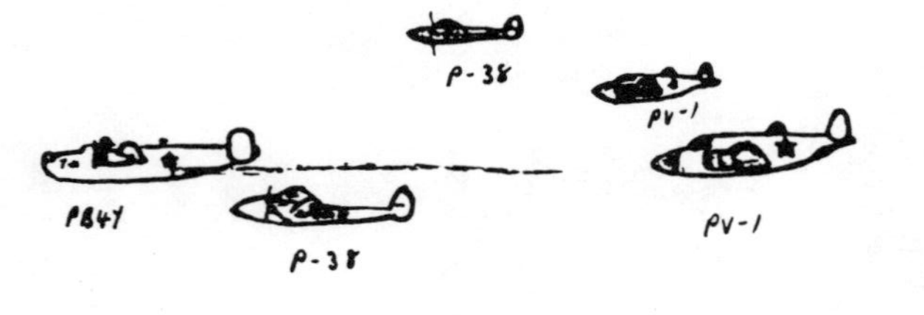

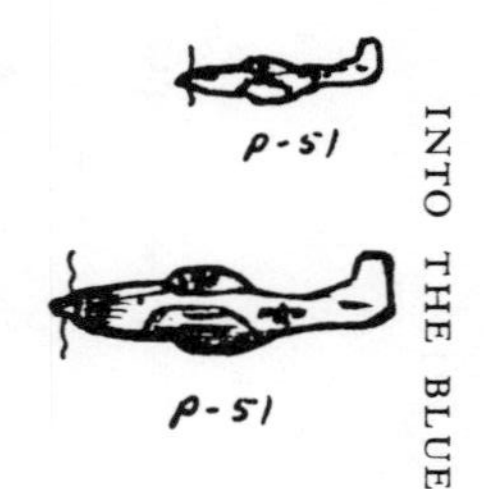

Formation of aircraft seen during early evening of 12 August 1944. Separation distances have been greatly protracted to conserve paper. Nearest aircraft to the Liberator at the time of the explosion was the P-38 to Port, also the Starboard PV-1 had closed on the Starboard P-38 shortly after passing the observer, and may have just overtaken it as the explosion occurred. The Two P-51s were catching up fast from the rear and may just be inquisitive pilots from Leiston having a closer look at the assorted group, as was a P-47 which passed directly overhead 'Dressers Cottage' at the same height and on the same course as the formation.

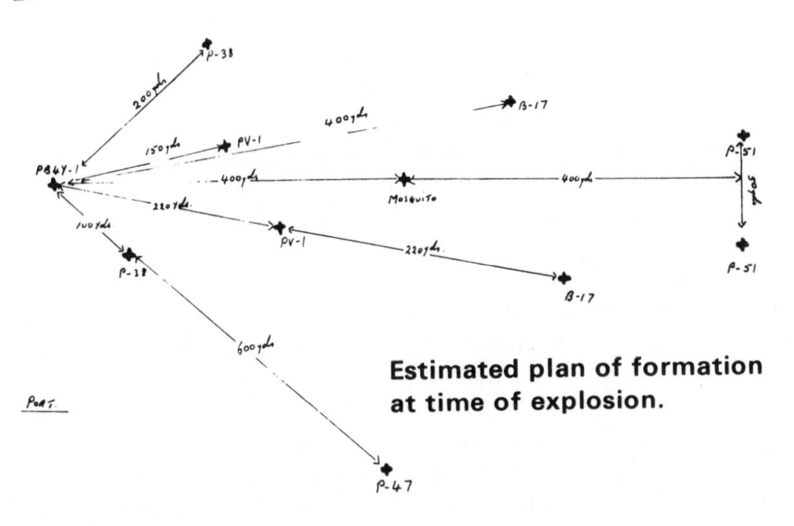

The Kennedy formation as remembered by Mike Muttitt.

and Willy would jump and be recovered by a waiting B-17 so slickly that the British would not have time to ask awkward questions about what they were doing. A gentle turn after that would head the robot out over the Channel for the target.

There was to be leisure to get the delicate job done right, to iron out any snags. No hurried baleouts here, the Navy would show the Army how. There was to be no unnecessary chatter to alert the Germans or distract attention. When the changeover of control had been satisfactorily achieved and checked, Kennedy was to broadcast the codewords '*Spade Flush*'. When their part of the arming procedure had been completed and Willy and Kennedy were ready to bale out, Kennedy's last words before leaving were to be '*Hearts Trump*'. From the television camera set up in the nose of the Liberator, the whole programme would be witnessed 'live' not only by the controllers in the mother-ships but by those with a receiver on the ground. Hopefully, they would have a close-up view of the V-weapon site's doors as the robot thundered towards them.

The crews of the two Venturas would not have been human if they had not felt some tension. Kennedy and Willy would bale out over England, but the control planes would have to go on to the French coast; their crews had been overseas only for a fortnight and had never been in action. Coolly, Kennedy completed the intricate hand-over procedure and the controller in the lead mother-ship, VK-13, *Zootsuit Red*, waited for his OK call before turning the drone left onto a course for Beccles. The time was now 6.15 pm.

'Spade Flush', came Kennedy's voice. And again, firmly, 'Spade Flush'.

The controller indicated a left turn to the drone and obediently the robot complied. All Kennedy and Willy had to do now was sit there and watch someone else flying their plane for them. The drone was roughly on course, but off to the right side (east) of its planned track, nearer to the sea than it should have been, and over Blythburgh. Still, there were seventy miles still to go before the two pilots had to bale out and leave it. Plenty of time for fine corrections to be made. Now it was 6.20 pm. Time on target was to be 7 o'clock precisely.

At 6.20, nine-year-old Michael Muttitt was with his brother in the garden of the house where they lived, Dressers Cottage, Darsham, when he heard the sound of a formation of aircraft approaching from the southeast. Mike was a keen aircraft recognition fan whose ambition was to become a pilot in the RAF. He had familiarised himself with most of the common Allied aircraft types by visits after school to a nearby USAAF base and by visiting sites of crashes and forced landings. A keen, technically proficient youngster always makes a good witness, certainly better than most adults. One remembered detail was to be crucial in trying to solve the mystery of what happened next. He wrote later:

> A loose gaggle of aircraft soon appeared led by a Liberator, followed by two P-38 Lightnings, two PV-1 Venturas (which at the time I misidentified as Hudsons), then a Mosquito and bringing up the rear two B-17 Flying Fortresses. A P-47 Thunderbolt flew directly over the house and two P-51 Mustangs were catching up the formation fast from their Leiston base.
>
> As the aircraft passed at about one mile to the East I could see a thin trail of smoke or vapour coming from the rear of the Liberator's weapon bay.

The deaths of Joe Kennedy and 'Bud' Willy as remembered by Mike Muttitt.

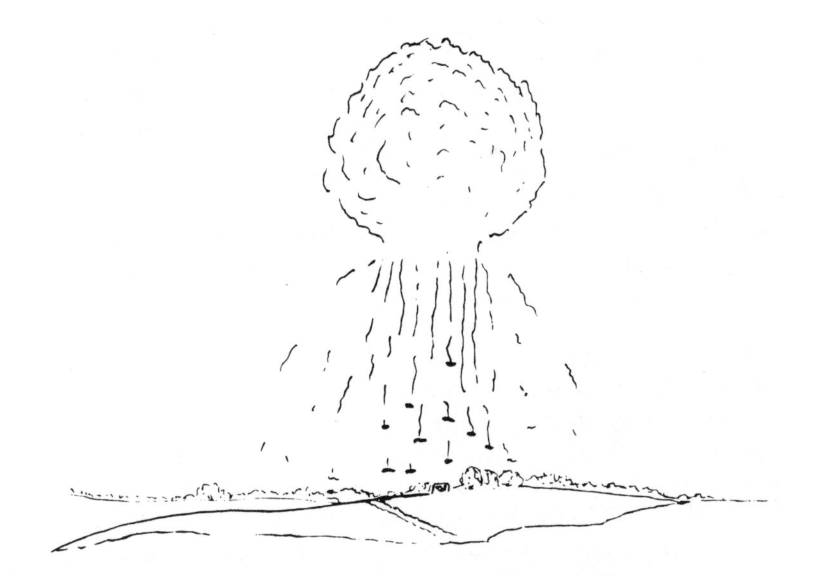

Pall of smoke a few seconds after the explosion with flaming wreckage already almost at ground level.

My eyes were still firmly fixed on this odd assortment of aircraft as it headed North towards Blythburgh when the lead aircraft suddenly disappeared in a giant ball of fire which rapidly changed to a swirling black mushroom-shaped pall of smoke with flaming wreckage plummeting earthwards and then the sound of the explosion reached us with an ear-shattering thunderclap (some observers reported a loud double explosion but from two miles distant it was far too deafening to my sensitive ears to discern).

This giant explosion brought my parents running into the garden and by this time all the wreckage had reached the ground and the enormous pall of black smoke had now changed its form to that of a giant octopus with ugly black tentacles formed by the trails of smoke left by the earthwards plunge of blazing wreckage.

Everything had happened so quickly that I had momentarily forgotten the other aircraft but can remember one of the Lightnings spinning earthwards (I heard that he recovered just in time over a hamlet three miles away) while other aircraft banked left and right away from the fireball. After a few minutes the only aircraft visible were the two Venturas circling the crash site, the pall of smoke still rapidly gaining altitude and drifting north-eastwards towards Southwold.

Robert B. Ball, crew chief and gunner in the back-up Ventura, was in the top turret looking forward, watching the red light mounted in the Liberator's tail as identification and guide to the mother-ships, when, without warning, the four-motor bomber 'literally disintegrated in a big ball of fire, followed by a cloud of black smoke.' The lead Ventura, closer to, felt the shock, and Colonel Elliott Roosevelt's photo Mosquito which had nosed in to 300 yards to get better pictures was damaged and the engineer-photographer hurt. One airman remembered seeing trees on the ground bend over from the blast wave, and then spring back. The presence of the Lightning which almost dived into the ground after the explosion led to a local rumour that it had shot down the Liberator.

Some people, who heard the bang from four miles away, took it for a German rocket exploding. One of them was Mr W.S. Jarvis, who was a full-time fireman at Southwold: 'We were working in the Fire Station yard, hanging up canvas hose to dry after attending a fire, when we heard a terrible explosion and saw a pall of smoke to the south-west of us. We notified the Watch Room and were ordered out to investigate.' After travelling six miles by road, they found an area of burning heathland called Blythburgh Fen, the nearest dwelling being the Game Keeper's lodge, New Delight Cottage, where relatives of Mike Muttit were living. 'The plane was scattered over a large area,' recollected Mr Jarvis, 'and we saw human remains, which were collected by the American Air Force fire and crash party we used to work with on these occasions.'

A police inspector from Southwold, Inspector Bird, went to the crash site with PC Roland Chipperfield from Blythburgh; it was said that they saw two torsos, an arm with a wristwatch, and enough for 'a boxful of meat'. The report from East Suffolk Police mentioned that 'small pieces of human flesh were found scattered' but 'no complete bodies were found.' Also, curiously, the police were 'unable to ascertain the number or home station of the plane'.

The cover-up had already started. Both the Army Air Corps and the US Navy were terrified of the Kennedy family finding out what had happened to their eldest son. When Admiral Samuel Eliot Morison, the noted American naval historian, came to write the story for the first time for *Look* magazine in 1962, he relied on official documents in Washington and was able to state that the explosion had been so complete that the Liberator had in effect vanished. 'When the ground was searched next day,' he wrote, 'nothing big enough to pick up was found – not even a button.'

No one could have checked him in 1962 because he gave the wrong crash site – 'a field near the village of St Margaret and St Lawrence' nowhere near Blythburgh Fen. This is where the Liberator should have been, but wasn't. The Admiral had also been misinformed by official documentation as to the target. He wrote: 'Helgoland, a proving ground for V2 missiles, was selected as target.' The 'red rock island' was in fact a U-boat base off the mouth of the Elbe in Northwest Germany, 350 miles away and in a different country from the real target, a V3 firing site in France. The Helgoland story has proved really lasting and is repeated to this day, on Morison's authority.

Part of the same deception is the appearance of Kennedy's name, and also that of Willy, on the wall of the 'missing' memorial at the US cemetery at Madingley in Cambridge. In East Anglia, however, there is a story that what was left of Kennedy was buried on an estate in Derbyshire belonging to a friend of the family.

At the time, and for a long time after, the locals did not know that the father and grandfather of airborne explosions had killed Joe Kennedy; most would not have known who he was in any case. None of them had been killed, but a large number of buildings had suffered minor damage, and the heath had been set on fire by blazing wreckage from the Liberator, parts of which lay scattered over a square mile of countryside.

Michael Muttitt visited the crash site next day:

> Wreckage varied in size from large pieces of wing spar and fuselage sections to small pieces no larger than a sixpence, the lighter pieces of metal appearing to have fallen almost directly below the explosion, starting the fire, while the much heavier

> engines, propellers and undercarriage assemblies had been carried north by not only the explosion but by the aircraft's momentum. I also vividly remember the fragments of parachute silk and parachute cord caught in the brambles alongside the B1125 road . . . Even when I joined the RAF in 1952 much of the wreckage was still lying around.

Christopher R. Elliott, who was also a schoolboy during the war, when the flatlands of East Anglia were virtually all one US aerodrome, photographed in 1947 'a badly damaged radial engine associated with the incident at Blythburgh'; it was only many years later, when he was a reporter on a local paper and his interest was sparked by Admiral Morison's misleading article in *Look*, that he undertook original research into wartime Police and ARP files, which showed that the Kennedy explosion had taken place fifteen miles from where Morison had placed it, and that the battered radial was from that particular Liberator.

Later, Stewart P. Evans, a young police officer interested in wartime crash sites, undertook more original research, especially into USAAF and US Navy records, and also produced a professional drawing of the distribution of the wreckage which Admiral Morison had been led to believe did not exist. Michael Muttitt also carried out original research, particularly after he had come to live in the gamekeeper's cottage which had been directly under the explosion and still showed damage caused by it. It was more than fifteen years after the war before these local people were able to start finding out what had happened on 12 August 1944.

The reason for this was the slickness of the cover-up operation. The official US Army Air Force crash report of 14 August 1944 was a marvel of apparent ignorance. Under the heading of 'unknown' was the serial number of the aircraft, the group to which it had belonged, the number and size of the bombs, and the cause of the accident. The report also mentioned that no notification of the accident had been made to RAF Bomb Disposal, Civilian Police, or RAF generally. The policy was silence. Admiral Morison could never have seen this, because the report correctly located the crash site, mentioned damage to at least sixty houses in Blythburgh and an unknown number of other places, and stated that one engine fell near Blythburgh Lodge, three engines near Hinton Lodge. So much for Morison's 'nothing was found – not so much as a button', which *Look* headlined as 'THE FIRST INSIDE STORY OF HOW JFK'S OLDER BROTHER MET HIS DEATH IN THE SKY.'

When Christopher Elliot wrote to Admiral Morison in 1973, he received a brief, stonewalling answer that he could not enlarge on the *Look* article, that he had his information from documents at the White House which Joe's brother, then President, had given him. And he concluded: 'Beyond that, I know nothing except that I had some knowledge of that country where he was shot down, and that was very little.' The reference to the Liberator having been 'shot down' seems an inexplicable lapse, like the mention of Helgoland as the target. But the two lapses do link, in that later drones did go to Helgoland and they were indeed all shot down.

In the immediate aftermath of the tragedy, anything seemed possible. Commander Smith even asked if either of the two mother-ships had pulled the 'Destruct' lever inadvertently; both proved to be still wired solid. First the junior officers of the unit and then Smith reviewed some likely causes and many unlikely. Unlikely were sabotage (plane well guarded and no German spies reported), static electricity (plane well grounded), petrol fumes leaking into bomb-bay and sparked (no one had smelled fumes in the plane), direct hit by British AA fire (no telltale puffs of flak seen by accompanying planes). Possible were instability of the Torpex explosive itself (actually a loader had dropped one of the boxes accidentally without blowing up the airfield, so this was unlikely), and the heating of electrical fuses.

It would be impolitic to admit any fault, so what was significant was what was actually done afterwards. The 'safety' pin was done away with: indeed there was to be no more electrical arming by remote control; arming was to be manual, carried out by the drone pilots themselves, just as the Army did it. Tests did seem to show that if the equipment received a signal by mistake from some outside source other than the mother-ships (which was quite possible), the restraining safety pin would prevent the charge from being armed but at the same time would cause the solenoids to heat up so much that in a few minutes they would blow the charge. If that theory was true, it was dynamite, so to speak. Representatives of Joe Kennedy Senior were sniffing round the base, trying to find out how Young Joe had died.

Mike Muttitt's observation of smoke or vapour trailing from

the bomb-bay of the Liberator, visible to a watcher on the ground but not necessarily from a plane above or behind, is some confirmation of the safety pin in the arming panel as the villain. Anyway, it was removed and there was no repetition of the Kennedy accident.

Of course, there was exaggeration. The 21,770 pounds of explosives in Kennedy's machine was claimed to be the largest charge ever detonated by man. But this figure must have been exceeded many times, right back to the days of gunpowder. At Messines Ridge in 1917, the British Army laid and detonated under the German defenders almost a million pounds of high explosive, in nineteen separate mines. And on Helgoland itself in 1947 the British Navy detonated (from a safe nine miles away) 6700 tons of unwanted high explosives. I saw the results of that myself and it was awe-inspiring. At Hanbury in Staffordshire there was also a 3000-ton RAF bomb dump which blew up in November 1944.

Kennedy's last words were the code 'Spade Flush', but his next-to-last words were often quoted, particularly at the time, when the jokes were funny. People remembered how he had donated his remaining stock of eggs (incredibly precious in wartime) to the other members of his hut if he didn't come back. A good joke then, but a bit flat now. And to someone who asked him if his insurance was paid up, he'd replied that no Kennedy had to worry about insurance. Today that sounds the boasting of a rich man, not the riposte of someone who stood to lose life itself within hours.

Commander Smith put Joe Kennedy in for the Congressional Medal of Honour, America's highest award, but he failed to get it. The medal he was given, posthumously, was the Navy Cross, so that in death he drew level with his brother John F. Kennedy in honour.

Much worse off was the girl he loved, whom he might have married had she not been already married. She had been expecting him for the weekend at Lady Sykes's estate in Yorkshire. Instead, she got the news of his death. Two days after that, came news of another death – her husband had been killed in Italy. Both men gone. That left just Kathleen Kennedy's husband, the Marquis of Hartington. Two weeks later he in turn was killed in Normandy. It was that sort of summer.

But Joe Kennedy's death was of a different dimension and raised wider issues than the merely technical reasons for it (few lamented

Willy, because his insistence on the retention of the safety pin may have been the immediate cause of both their deaths and the ruin of the mission). Could the mission have been successful? Was the whole thing – the Aphrodite, Double Azon, Castor and Anvil projects combined – a practical, necessary, and even vital operation of war, as its leaders had claimed and its crews had been told? It is not necessary to purloin top secret files from the Pentagon to be sure it was nothing of the sort.

One's own experience, shared by many, was that Hitler's V-weapons were not effective, not even very damaging to morale, because they could be aimed only at areas, not at worthwhile targets. They did, however, signal a change in warfare. I wrote in my diary on 17 June 1944, in London: 'Weird, this menace from a piece of mechanism. Doesn't seem like an air raid at all. Like a story about the future, by some fantastic author, but more eerie and uncanny.'

On 26 July I left London for Normandy to join the First Canadian Army, whose task was to advance left of the line and then clear what we called the 'Rocket Coast' of the Pas de Calais. It was not until the end of the month that the US Navy's Anvil personnel were all gathered at Fersfield, and it was not until the first week of August that the Army Air Corps began flying missions against the V-bomb sites. Almost exactly a month later, on 7 September, I was in the Pas de Calais looking at the latest type of V1 site. I wrote in my diary: 'Never was an anti-climax so complete. I had expected, from the newspaper reports, to see some colossal concrete erection, some giant tunnel, containing elaborate design and complex gadgets; and this might have been a house in course of construction, so simple was it.' It was just two wedge-shaped ramps, aligned on London, emplaced at the edge of a wood and flanked by half-sunken concrete blockhouses. It had been very thoroughly bombed, conventionally, to not very great effect. I noted that whereas we called these things "Doodlebugs' or 'Buzz-Bombs', the local French said 'Météor' or 'Forpille'.

However, the great concrete structures did exist, but there were not many of them and they had been most thoroughly plastered by both conventional bombing and also by the new weapon developed for the RAF, the 6-ton 'Tallboy' high penetration bomb, which if it struck earth would go so deep that it would create a giant sub-surface hole most damaging to foundations. I saw one of the giant sites which had been attacked both conventionally and by 617 Squadron using 'Tallboys', and noted in my diary:

> We came down to a small town lying at the foot of quarried hills. Ninety per cent of the houses in the town had been damaged by a blanket of bombs laid upon the countryside and cutting the road in half in half-a-dozen places. Slateless roofs lay flat on the rubble of walls that had crumbled to dust, children walked in buildings blown open to the winds, walls of houses sagged and bulged, with great cracks crawling crazily across them, the church steeple rose brokenly, its spire snapped off and its sides torn open. And still we got a few waves and smiles (from the inhabitants) as we passed. Black-and-orange German signs proclaimed that the poor ruin was Wizernes.

All these captured sites were inspected shortly afterwards by British experts of the Sanders Mission. Wizernes had been attacked conventionally from March 1944 onwards; and on 17 July by 617 Squadron, which caused the Germans to abandon the site, intended for the launch of V2 rockets. A similar V2 launch site at Watten had been attacked as early as 27 August 1943 by 183 Flying Fortresses during the critical period when the fresh concrete was setting; this conventional attack had been totally effective. But the Germans started to erect new constructions there, which 617 caused them to abandon by an attack on 25 July 1944. The Siracourt and Lottinghem V1 storage and launch sites had not been completed when captured, but bombing of both by conventional means and of Siracourt additionally by the special weapons of 617 Squadron, had not been very effective.

The Kennedy target at Marquise Mimoyecques, ninety-five miles from central London, proved to be an emplacement for V3 rocket guns, something not known until after capture. Heavy raids by RAF Bomber Command on 19 and 26 March did not stop the German construction work, but a raid by 617 Squadron with 'Tallboys' on 6 July 1944, scoring only one direct hit, nevertheless collapsed one gun shaft and some tunnels by the 'earthquake' effect of four near-misses. This damage was irreparable by the Germans in the brief time allowed them before the Allied ground forces overran all the sites. Third Canadian Infantry Division had reached Guines near Calais, a few miles from Mimoyecques, by 5 September, and although the German fortresses of Boulogne, Gris

Nez and Calais held out against us two or three weeks more, the German V-weapon threat had been over for forty-eight hours by the time I saw my first site on 7 September. The Kennedy mission had become totally redundant just twenty-four days after the Liberator had exploded.

None of the large construction sites fired a shot: no V2s or V3s were ever launched from the Pas de Calais, although many thousands of V1s came buzzing away from the simpler, cheaper launch sites. However, Mimoyecques was to have one great distinction beyond all the others: it was the only V-weapon site intact and dangerous enough for Colonel Sanders to recommend, on 21 February 1945, that it be demolished as a permanent potential menace to London. French permission was sought but not obtained, so the British blew a ten-ton charge anyway on 9 May, which proved unequal to the task before it; a further twenty-five tons had to be exploded at the railway tunnel entrances on 14 May, before London could be declared safe. So even if the American drone had hit an entrance, it is by no means sure that its twelve tons of Torpex and TNT would have sufficed.

It is clear therefore that no immediate V-weapon threat existed when the first futile Aphrodite and Anvil attacks were launched in August 1944; the sites had been crippled already, largely by the use of the Barnes Wallis 6-ton 'Tallboy' penetration bomb delivered with the required great accuracy by 617 Squadron of the RAF on various dates in July. The threat was abolished completely when First Canadian Army captured the 'Rocket Coast' in the first week of September. But one must go on to ask: could the drones ever have done the task, had the requirement remained? Were they in fact, as General Arnold believed, the face of the future?

The brief answer is: No. The future lay in the development of the V1 and the V2 launched from inconspicuous or mobile sites, rather than the concrete monstrosities which had served the Germans so ill in France. I was able to note that next stage also, firstly on the faster-than-sound receiving end of the first V2s into Antwerp in October, and later from Nijmegen I could watch the V2 firings from somewhere near Arnhem, the visual sign being an almost vertical contrail streaking up from the ground in daylight before arching over for Antwerp and London. That indeed was the future.

No one could stop a V2 then. In contrast, what chance did the

US Navy and Army drones have? They were big, they were slow, they were low; and they required good weather for visual control by the mother-ships. But the Navy tried once more. Commander Smith would not rest content with the Kennedy mission. On 3 September, a war-weary USAAF Liberator was sent out, was armed manually, the lone pilot H.R. Spalding baled out, and the drone was aimed at the impregnable U-boat pens in the harbour of the rock island of Helgoland. The operator picked the wrong harbour, the breakwater of the nearby sand island of Dune, but it made little difference, as with eight seconds to go before impact the TV picture was lost, indicating that the drone had been hit by flak. Damage to barrack buildings was claimed, but it seems what was hit was a coal dump. Well, coal is a war-winning substance, but it seemed an expensive way of scattering the stuff. After that, the Navy's effort was wound up, but the Army continued.

Helgoland continued to be a target but the U-boat pens were not hit and no military damage was done; one pilot lost his life baling out of a B17 robot. Another drone hit, not Helgoland but Sweden, so the Swedes added an American secret weapon to their collection of V1 and V2 remains. Finally, in despair, abandoning key targets altogether, the drones joined RAF Bomber Command in indiscriminate attacks on built-up areas, the least productive and most inhumane method of all. At least they wouldn't lose any aircrews. On 1 January 1945, two robots were sent to the town of Oldenburg in Northwest Germany – one missed by two miles, the other by five.

The record stood at two attacks by the US Navy – both unsuccessful; and seventeen attacks by the US Army Air Corps – all unsuccessful. Nineteen robots, nineteen failures. Four deaths and a number of injuries for nothing. Aphrodite was wound up, and on the last page of the last document, some disillusioned Yank had written:

'THE END – we hope.'

Chapter 13

BACK FROM THE BED OF THE SEA
Zuyder Zee: 1917–1950

Probably the dearest dream of every small boy is to get hold of a machinegun for his very own and enough ammunition to blaze away for hours. Without intent to harm anything, naturally, just to experience the noise for its own sake. Better still would be a complete gun turret fully equipped with multiple machineguns, with which to have an exciting make-believe battle to rival the one in *Star Wars*. In 1960, that is exactly what a group of Dutch boys, the eldest aged fifteen, actually did. However, the sounds of firing out on the marshes near Kampen, on land recently reclaimed from the Zuyder Zee, was reported to the police, and the Royal Netherlands Air Force was called in.

The scene was the new polder Oostelijk Flevoland, which had been enclosed and the sea pumped out during 1957 and 1958, as part of a vast land reclamation project. It was expected that the remains of long-sunken ships would be found, but the authorities did not at first realise the strange harvest of lost aircraft and missing men which lay hidden among the wilderness of muddy reeds which appeared after the sea water had been gradually pumped out. The first discovery had been of a Gloster Meteor VII jet fighter of the Dutch air force, which had gone down into the shallow sea, only twelve to fifteen feet deep, in 1950 and vanished. While this aircraft was being salvaged by a detachment from the air force in 1960, four other planes were found close by. Two were fighters, a Spitfire and a Messerschmitt 109; two were four-motor bombers, a British Lancaster and an American B17 Flying Fortress (with ninety-four small bombs still aboard). And then came the stories of wild firing going on out on the marshes near the Kamperdijk, the polder's northern dyke.

The newly-born polder was hard to penetrate: until it dried it was a kind of muddy no-man's-land, mysterious and lonely.

Stumbling into it, the boys had come across an object like giant clam – the rear turret of a disintegrated bomber, complete with two Browning machineguns. Over a period of weeks, the boys hacksawed the turret from the rest of the fuselage and took it away on a sled; then they cleaned the guns, collected ammunition from the wreck, mounted the turret on a wooden base – and were ready for firing practice which made imitation Cowboys 'n Injuns seem tame. After all, whoever heard of a hero of the Wild West getting 1260 rounds a minute out of his six-shooter?

When the air force team reached the crash site, they found two radial engines fitted with three-bladed wooden propellers amid the wreckage, clearly that of a Wellington bomber, and one of the engines had a number. In spite of that, identification was not possible then. Indeed, some people took the view that trying to identify such remains was like trying to empty the sea out of the oceans with a teaspoon. Others believed that they had a responsibility to the relatives of the crews to provide identification. As more and more human remains turned up in and around the lost planes, it was the second view which prevailed. If the depth of burial was shallow, little or nothing remained except bones; but where the aircraft had gone deep into the seabed, twelve feet down or more, and especially where it had entered a layer of clay which had closed in above it, then often the excavators found 'fresh' flesh, muscles, nails, hair, and so on. At first, they flinched at the macabre finds, but soon got used to it and, even if they had not liked the task to start with, became fascinated by the 'detective' aspect of the work, so that each new wreck was like finding gold.

By this time a special unit of the air force had been set up to deal with what was now regarded as an historical study. It was commanded by Lieutenant-Colonel A.P. de Jong. During the war, the Zuyder Zee's vast expanses of shallow water had in effect provided a flak-free entry into the heart of Europe for Allied bombers. Then to counter this, the Germans had set up a tight network of fighter defences, so that many air battles took place above it, the victims going down into the waves to vanish apparently forever under the sea. Exact figures were impossible to come by, and estimates of about 1000 British, American and German aircraft were arrived at as a rough approximation. As by no means all the Zuyder Zee had been reclaimed, the air force unit worked in close cooperation with the Ministry of Waterways.

There was an economic reason here, for any fisherman who could prove that the object which had torn his net was either a sunken ship or a crashed aircraft could claim compensation from the Government. The necessary diving and lifting was carried out by a cable-laying ship, the *Poolster*.

Some very curious remains turned up. In 1953, the *Poolster* lifted an obstruction which proved to be a strange, yellow-painted aircraft. One engine had odd writing on it, which no one could read, and was in fact Cyrillic (i.e., Russian) script, but the machine had Luftwaffe black crosses painted on wings and fuselage. It proved to be a Tupolev SB-2 light bomber captured by the Germans and used by the Luftwaffe for training and target towing because they were so slow and old-fashioned. Then there was a British Mosquito, or parts of it, in association with German machine-gun ammunition – perhaps a captured machine flown by KG 200, a special unit which used captured Allied aircraft operationally, particularly for reconnaissance. From the centre of what was left of the Zuyder Zee (now called the Isselmeer) came a V2 rocket and motor, probably fired from Hellendoorn, sixty miles away. Of the same vintage was a rare bird – a Messerschmitt 262 jet fighter. With all these, found in the 1950s, no great trouble was taken; they were just removed as obstructive scrap metal. In the 1960s, the real detective work began, with the relatives in mind.

On 26 July 1963, for instance, the remains of a large aircraft were discovered broken into small pieces and scattered over an area measuring 50 × 50 yards. What type of aircraft? What nationality? The bits varied in size from one inch to five foot long. One of them had an RAF roundel painted on it. Parts of the fin and rudder identified the type – Lancaster. Then a Merlin engine with a readable serial number – MXX 100597. After a long search by the makers, A.V. Roe and Co, the particular machine was traced – it belonged to No. 12 Squadron, missing from its thirty-seventh mission on the night of 11/12 June 1943. That night, the Squadron lost four Lancasters, and there were no survivors from any of them. Research in the Netherlands showed that five bodies from this particular aircraft had been washed ashore and buried, but there was no sign of the remaining crew members in the wreckage, probably because the Lancaster had broken up in the air after an attack by a German night-fighter. One of the engines was not found either, but one of the recovered propellers was used as part of a memorial set up at Dronton to all airmen who died over Holland during the war.

During the summer of 1963, while the Excavation Service was busy reclaiming tons of scrap metal from this machine, a labourer drove past them on his tractor. 'Say, fellows,' he called out, 'if you've got a minute to spare, have a look at Lot S-54. I found some old, rusty bicycle-type wheels on a wide axle over there. It looks like a real old-fashioned aeroplane. And there's more of it buried in the ground.'

Up to this time, the excavators had never seen a really old wreck, although many odd types had turned up. Excitedly, they went over to look, and saw at once that this was no Second World War crash. There was the bicycle-type landing gear and ten yards away was an old honeycomb radiator for a Mercedes engine, such as the famous Fokker DVII single-seat fighter had used. Among the pieces were fragments of copper, showing that the plane had belonged to the pre-aluminium age.

Next day the excavation proper began. As the mud was removed, new finds were made. A heavy copper fuel tank. A smashed instrument panel, with a rev-counter measuring up to 1800 revs/min., made in Leipzig. Then a practically undamaged machinegun which when cleaned proved to bear the inscription 'Parabellum, Berlin, 1913'. Then a circular drum holding dozens of 7.7 mm cartridges in excellent condition; and each one had '1916' marked on it. Next, an elevator and part of the wing centre-section – all painted khaki-green. After this, a small pear-shaped 25-pound bomb with a single central fin; a Heath Robinson bombsight with three lenses; flying wires, tubes and even aircraft tyres with the rubber in amazingly good condition. And a flare-pistol with a big wooden grip and four red flares to go with it. All these were obviously destined for their own museum; but what type of aircraft was it?

A number of handbooks were consulted and the vital information proved to consist of the shape and size of the landing gear, the diameter of the wheels and the dimensions of the axle. It was a Gotha, either a G-IV or a G-V. A long-range heavy bomber of the *Englandgeschwader*, the unit set up to carry on the bombing of London after the Zeppelins had been beaten, which was in its turn to be mastered by the defences.

That narrowed the timespan a little, to around 1917–1918. In official records mention was found of a Gotha which crashed south of Urk, and the date of the log entry was 13 October 1917. The 'detectives' then turned to the files of the local newspaper, the *Urker Courant*, of around that date, and found the developments in the story reported on 13, 20, and 27 October. The first item had merely reported the finding of a body by a shrimp fisherman, Jan H. Wakker, captain of the UK-125. A watch with radium hands and a purse containing a few Marks was found on the corpse, which was established to be that of Lieutenant Martin Emmler of Ober Düsseldorf. Speculation connected this body with a new obstruction in the Zuyder Zee which had damaged the nets of an Elburg fisherman earlier that week.

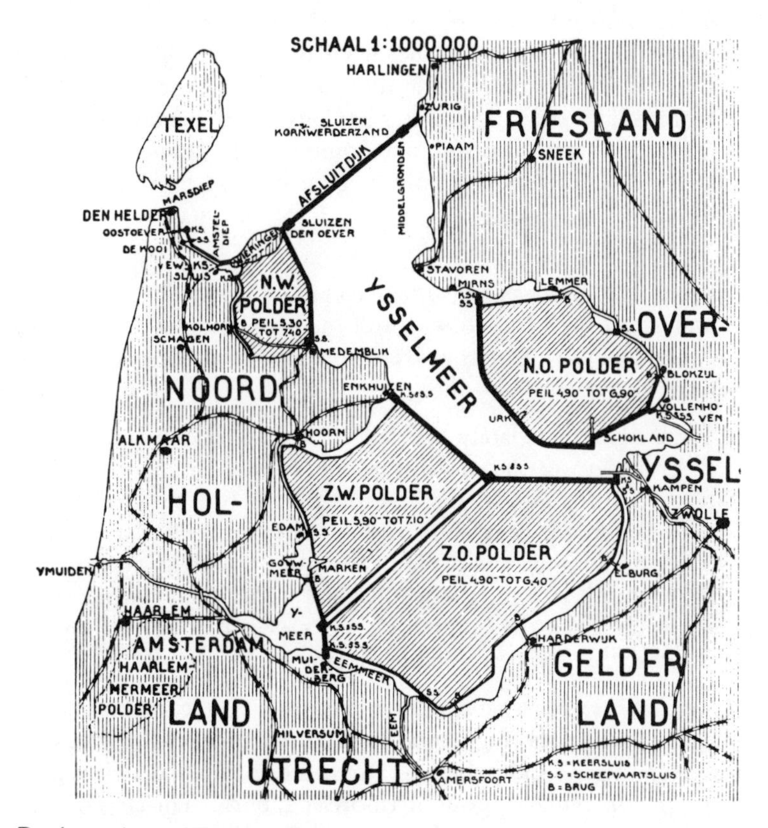

Reclamation of Zuyder Zee

When, half a century afterwards, the Dutch air force excavators took up the trail, it was not quite cold; in the end they were able to contact seventy-seven-year-old Hessel Wakker, the son of Jan H. Wakker, who had been aboard his father's shrimp boat that day in October 1917. His story was this:

> I was sailing with my brother on my father's fishing boat by then. We were fishing shrimps in the neighbourhood of Schokland Island and had our nets in the water and were drinking a cup of tea in the boat's cabin. Coming on deck, we saw an object adrift quite near our boat, but we couldn't make out what it was. Then, as it drifted alongside, we saw that it was a clothed human body, floating head-down. I grabbed it by the collar and with the help of my brother managed to bring it on deck. I remember that the man had a wadded coat and had been injured on the front of his head. We brought the body ashore at Urk where the municipal authorities took charge of it. A few months later the burgomaster gave us 25 guilders, as a reward from the German consul, and some months later we received another 25 guilders, this time from the pilot's family. We mostly earned 10 to 15 guilders a week with fishing, so this reward was very welcome.

From the local paper, the excavators learned that Martin Emmler had been buried with full military honours at Kampen, and that the body of another German airman had been found. The newspaper report went on:

> Some fishermen lately spoiled their nets on an underwater obstruction to the northwest of Elburg in the so-called Schokker channel. After investigation it became clear that a giant aeroplane was sunk there, approximately sixty metres long. An attempt was made to bring it up with an anchor, but the anchor broke and the whole giant went to the bottom of the sea again. Only the bomb-rack was saved. It is not yet known whether this twin-engined aeroplane carried bombs.

Yet another newspaper report told the 1963 excavation why they had found only fragments of the Gotha. Most of the wreckage had been raised in October 1917 by a steamship and two fishing boats and the remains brought ashore at Elburg. 'The aircraft, one of the newest big fighting planes with two enormous engines, was completely disintegrated. Some bullets also were found. It is thought that a machinegun still lies at the bottom of the sea.'

This information fitted perfectly with the single Mercedes engine found in 1963, together with a single Parabellum machinegun. There should have been a third man in the crew, and the excavators had indeed found human remains among the wreckage, although these consisted only of small pieces of pulverised bone, none more than an inch long. As the deepest part of the Gotha was buried only three feet under the soil and the whole area had been dried out four years earlier, it was not surprising that so little was left that the third body was unidentifiable. The facts also fitted the three brief log entries on the incident made by the Dutch military aviation authorities in October 1917. The aircraft remains had been taken to the Soesterberg base, but nothing could be salvaged for Dutch use. The excavators noted that this was the only report of a German crash in the Zuyder Zee during the First World War which had been mentioned in the Dutch air force daily official logbook.

In March 1969, six years after the excavation, the case ceased to be of academic interest only, Herr Ernst Pfort of Baden-Baden wrote to the excavation unit of the Dutch air force, saying that he had read an article in the *Badische Tagblatt* of 10 August 1963 about the recovery of a Fokker DVII which had crashed in the Zuyder Zee on 13 October 1917, perhaps after making a bombing raid on Paris or London. Herr Pfort recalled that his brother, Karl Pfort, had been a lieutenant in command of a Staffel of *Bombengeschwader* 3, the so-called '*Englandgeschwader*', and he had not returned from a mission on the night of 12/13 October 1918. Could there possibly be any connection?

The 'Fokker DVII' story had been a journalist's first assumption on the finding of a *single* Mercedes engine; it took no account of the size of the wheels, which were for a vastly larger aircraft than a single-seater scout. So that was a piece of nonsense which could be put in the wastepaper basket. An odd coincidence was the date given by Ernest Pfort of 13 October 1918, with the Dutch log report of 13 October 1917 – exactly one year's difference! Quite an easy slip to make, on the part of Herr Ernst Pfort. But it wasn't! On looking carefully at the reports of the *Urker Courant*, it was clear that the first was dated 13 October 1917, but the inner logic of the story suggested that the aircraft had crashed earlier, having snagged nets some days before the story was printed.

That fitted a period of intensive Gotha activity over London during September and early October 1917. The Dutch looked in Morris's book *The German Air Raids on Great Britain: 1914–1918*, and found that a Gotha went missing from the raid of 25/26 September. This was also mentioned in a book I have in my library, Air Commodore Charlton's *War Over England*, which stated: 'It is possible that one Gotha which failed to get back had been damaged in aerial combat previously and fell into the sea.' The Dutch were understandably not keen about a date late in 1918, as they could not discover that the Germans carried out any Gotha raids after July 1918, but because German records were not complete it was impossible to be categoric about this. They could say however that there was no raid on the night of 12/13 October 1917. The 25th/26th of September or the 28th/29th of September were more likely dates; the latter night was very cloudy and a lost Gotha straggling back from England might well have violated neutral Dutch air space and vanished into the Zuyder Zee. The pitiful, pulverised bones might have been all that remained of Karl Pfort, but his brother would never know for sure.

Many a detective inquiry ended like that, but not all. Some were easy. In 1961, an Me 109G-6 of 5 Staffel JG 1 was identified by structural parts remaining at the crash site near Biddinghuizen in East Flevoland, and the pilot by the finding of his identity discs. He was Unteroffizier (Senior NCO) Fritz Kostenbader, who had previously been listed as missing after a vicious fighter v. fighter battle on 29 November 1943. Seven Messerschmitts were lost, while their first opponents, the P-38s of the USAAF 55th Fighter Group, lost eight. P-47s of the 56th Fighter Group were also engaged, but Kostenbader probably fell victim to a P-38.

In 1963 the German magazine *Jägerblatt* published an article on the excavation by the Dutch air force of Kostenbader and his aircraft. Colonel de Jong recalled the result.

> Soon after this publication I received a number of inquiries from relatives of German pilots still listed as missing over Holland. Among them was a telephone call from Frau Willius, the widow of an Oberleutnant Willius, a Focke-Wulf pilot. As up to that time we had still to find an FW-190 at all, we were unable to help her. In November 1966 we had a letter also inquiring about Willius, giving all known details of his last flight, and asking us again to try to discover his fate. We did not have a single clue towards solving this problem.

Identifying an aircraft which has already been found is one thing; to find a particular aircraft on demand, which might be almost anywhere in the Zuyder Zee area, and in any case could be completely buried under the earth and invisible, or lying in or under the bed of the sea, is quite another. Nevertheless it was done.

A helpful factor was that Oberleutnant Karl-Heinz Willius was a well-known squadron leader with forty-seven 'kills' who had been decorated with the Knight's Cross. On the day of his death he commanded 2 Staffel of JG 26 (the 'Abbeville Boys'), stationed at Wevelghem in Belgium. His number two, Leutnant Schild, survived to be debriefed, and that report was later published by Joseph Priller in his book *Geschichte eines Jagdgeschwaders, JG26.*

The day was 8 April 1944, and the operation was yet another attempt to cripple the Luftwaffe in advance of D-Day in the West. At 13.40 hours the first of 500 four-motor bombers forming a pathway a hundred miles or so long crossed the Dutch coast bound for aircraft factories near Brunswick and fighter airfields around the Ruhr. Colonel de Jong was living in Rotterdam at that time and saw them go over, very high up, heading east.

JG 26 scrambled at 14.15 hours. Willius led his sub-unit of ten FW-190s off the Belgian airfield and was directed by radio to the Zuyder Zee at 24,000 feet. He sighted a formation of three dozen B-24 Liberators and attacked them head-on with his ten Focke-Wulfs, taking the lead machine as his personal target. It crashed in the Noordoostpolder, all but one of the ten-man crew escaping.

As the Germans broke away to the south they were jumped by a group of high-flying P-47 (Thunderbolt) escort fighters, diving out of the sun. Schild, flying directly behind his leader, reported:

> I saw pieces come off Willius' fighter, which left a long trail of fuel leaking from the starboard wing. At that moment I had a P-47 on my tail and tracers were flying left and right along my cockpit. I flew behind 'Charly' (Willius) and attacked the P-47 which was sitting on his tail. It was a deflection shot but I saw the Yank Thunderbolt going down with a black trail of smoke; probably I destroyed it. This dogfight took place at 24,000 feet between Zwolle and Meppel and I believe that 'Charly', if he did not succeed in leaving his damaged aircraft, crashed on the eastern banks of the Zuyder Zee. He is still missing.

Schild also reported the loss of another FW-190 in the same area, in which Unterofizier Emil Babenz was definitely killed.

The Dutch were able to establish that three aircraft came down almost simultaneously at 15.00 hours: the leading B-24 of the formation attacked by Willius, Willius himself, and a Thunderbolt piloted by J.B. Dickson, who was killed and buried at Westellingwerf.

In October 1967, a year after receiving the last letter from Germany inquiring about Willius, the excavation unit was in the process of identifying a British Short Stirling four-engined bomber of No. 15 Squadron, when a reporter from the daily newspaper *De Telegraaf* visited them. His article brought the editor a letter from a farmer living at Kamperzeedijk:

> An aircraft crashed about one hundred yards from our farmhouse during the war! We believe the pilot is still in the wreckage. We would be happy to get in touch with the family of this man, but our efforts have been in vain so far. This aircraft came down on a Saturday afternoon, the day before Easter 1944, 8th April, at three o'clock. It was a German aircraft with one crew member. The Germans went to the crater in my meadow but made no effort to excavate; they only filled the hole in the ground. They mentioned then that it was a Focke-Wulf aircraft. We would be very much obliged if you could find an authority who deals with these affairs.

Colonel de Jong was elated:

> The answer to the jigsaw puzzle suddenly seemed to be found. Many parts of both stories fitted into each other: the date, the time of the day, the aircraft type, the location, the fact that the pilot was still unidentified. I decided to contact the farmer, Jan Kloosterman, the next morning. The farm proved to be less than a mile to the east of the Ijsselmeer dike. Mr Kloosterman had been an eye-witness and he remembered the details remarkably well because the day of the crash was also his wife's birthday. He described the dogfight with the '4-mots', seeing one bomber going down to the north when suddenly the fighter came down with terrific speed, seemingly aimed at their house. It was pouring smoke from the right wing, the propeller was idling and the canopy was still closed on impact. His story fitted completely with Lt Schild's official report. We tried out a bomb-locator on the site and this indicated that the wreckage was deeply buried, some three or four yards below ground.

When the salvage team arrived to begin the excavation, Colonel de Jong told them that the primary object in this case was to find and identify human remains, so great care and accuracy would be needed. After fifteen minutes' digging, pieces of metal were uncovered; then engine fragments, the dented propeller, 20-mm cartridges and armour plate. By now, it was clear that this was indeed an FW-190. As work went on, a plate bearing the aircraft's serial number 170009 was found; that was conclusive – this was Willius' aircraft. Then a shoe was unearthed and pieces of a jacket; finally, human bones came to light.

The remains of the pilot were removed to the German military cemetery at Ysselsteyn, where some 32,000 Germans killed in Holland during the Second World War are buried. The widow, Frau Willius-Cools, was informed and soon after visited both the crash site where her husband died and the cemetery at Ysselsteyn, accompanied by her daughter. At last, after so many years, she was able to place flowers on her husband's grave.

The finding of other aircraft sometimes helped to piece together an otherwise broken story, even when the aircraft had disintegrated. This was the case with two B-17 Flying Fortresses of the Eighth Air Force, shot down on the same day, 11 January 1944. For two months bad weather had prevented high-level daylight precision bombing of vital targets by the Americans. Blind bombing of ports through cloud was no substitute for the accurate attacks on German aircraft factories and airfields which were necessary to cripple the Luftwaffe before the invasion planned for May. And on 11 January, the weather cleared. 633 heavy four-motor bombers took off in three separate formations and headed for Holland, escorted by eleven groups of Thunderbolts, two groups of Lightnings (P-38s), six groups of British Spitfires and one squadron of Mustangs (P-51s). Only the Mustangs, the new long-range American fighters, could go with the bombers all the way to the target; but there were only forty-nine of these new machines available. Things started to go wrong almost at once.

The weather over England worsened, and the bombers took a long time getting off and forming up; as a result the fighters were forced to use up some of their precious fuel to no purpose. Then the weather over Germany worsened, and the Americans ordered back the second and third bomber formations as they were crossing the German frontier, leaving only the first formation of 238 heavies escorted by forty-nine Mustangs to press on alone a further sixty to one hundred miles into Germany. The course of the bombers towards the targets in the areas of Brunswick, Halberstadt and Aschersleben was almost the course for Berlin, and the Germans reacted with energy and determination. Fighters swarmed into the air from Döberitz near Berlin, Stade on the Elbe near Hamburg, and Deelen near Arnhem in Holland, and in spite of the cloud 207 out of the 239 scrambled made contact with the 238 bombers and their forty-nine-strong escort. The skies spewed stricken four-motor bombers and falling fighters.

The Germans lost thirty-nine fighters and the Americans claimed to have shot down 152. The Americans lost sixty big bombers and five fighters, and the Germans at first claimed 105 bombers. These overclaims were because several fighters usually attacked one bomber and many American gunners fired on each German fighter; also, they were very young men, and very excited, and things were happening with unbelievable swiftness. For some, that is. Other bombers took a long time to die, as examination of the wreckage many years later revealed.

From one B-17 of 533 Squadron from Ridgewell there was only one survivor, the left waist gunner, Sergeant John R. Lantz, in spite of the fact that he saw four other men on parachutes coming down and his last sight of the remaining men in the machine had shown them alive and unwounded. This particular Fortress had aborted its mission at the German border and been engaged by German fighters as it returned over the Zuyder Zee at 21,000 feet, exploding shortly after Lantz left it, or so he thought.

The wreckage found in 1966 near Urk was easily identified and told a grimmer story. Virtually everything found was from the fuselage, not from the wings: and the fuselage had not disintegrated until it had hit the water. It followed that the wings had come off, leaving the body of the machine with the remaining crew members alive and well to fall sheer from 21,000 feet without hope of escape, trapped inside their thin shell of metal.

Five of the Americans were reported killed but four are still listed as 'missing in action, believed killed'.

What puzzled the Dutch about this crash was the number of shoes found in the wreckage. Everything else was straightforward. The fin had been salvaged and that was marked with the Squadron insignia; a brass plate with the serial number of the aircraft was also turned up. But the shoes! So many pairs of shoes – why? What did it mean? Later, the Dutch began to realise that almost all American bombers showed this characteristic. The answer proved to be simple. As floppy flying-boots are awkward to walk in, the men had brought along a pair of walking shoes apiece in case of a forced landing or a bale-out, but almost never did they have the chance to put them on.

The other wreck from this raid which told a story was a similar B-17 of 812 Squadron from Alconbury. Dutch eye-witnesses in the town of Harderwijk saw it go down into the Ijsselmeer five miles away, under attack by German fighters. Four parachutes blossomed. One American came down safely on land, three fell into the waters of the Ijsselmeer. Fishermen reached two of them in time, but the fourth airman died of shock and cold as he was being picked up from the icy sea. The three survivors were Sergeants Glenn, Mills and Rajala. Later, six bodies were washed ashore at various points along the coast, with one man missing. Examination of the wreckage, which was complete, although in pieces, testified to a long and bitterly-fought retirement from the target, Halberstadt. The other machines of the formation had their last sight of this machine, lagging behind and under attack by fighters near Hallendorf, 202 miles away from where it at length crashed into the Zuyder Zee. When the remains were salvaged in 1969 it was discovered that three engines had been running at the time of the crash, but the fourth had been stopped. The body of the missing co-pilot lay a short distance away. It was easy to reconstruct the last one and a half hours of the dead men's lives. At 11.58 hours the bomber had been crippled over Hallendorf, and at 13.30 hours it had gone down into the Zuyder Zee, having fought a running battle for two hundred miles on three engines.

A surprising discovery in this B-17 was a badly-damaged aerial camera, far from waterproof, which contained film capable of being developed and printed after a quarter of a century under water.

Almost as surprising was a half-filled bottle of Eau-de-Cologne found in the wreck of a Junkers Ju 88D-2 which had crashed in an 80-degree dive. The scent was still recognisable. In a sense, as Colonel de Jong points out, burial – particularly in clay – serves to 'mummify' aircraft and their contents. Nowadays, some of those machines are genuine museum pieces, historically valuable.

Less welcome substances also survive. Many a laugh could be had in England during the war by pondering the fate of some land speculator of the future who would build his house on top of a big unexploded bomb. Worse very nearly happened to a Dutch farmer who wanted to build a house for his son next door to the farm on an open space where a Wellington had crashed in June 1941, shot down by a night-fighter. Only one man had been thrown clear and formally buried. The rest of the crew were thirty feet underground with one 1000-lb bomb, four 500-lb bombs, and a number of propaganda leaflets. Only the latter were now ineffective. As the Dutch method of building on such soil involves driving long concrete posts deep into the ground, the farmer was lucky that a salvage team was called in to excavate first.

There are two diametrically opposed ways of looking at the recovery and identification (if possible) of war dead. Some relatives, probably the majority, are glad to have the solution to a mystery; to substitute a known grave for all kinds of speculation as to what the term 'missing in action, believed killed' may mean. Others perhaps, those who hope against hope even after twenty, thirty, forty years, that their loved one has somehow survived, may prefer not to have the dream shattered.

One point may be cleared up. The readiness of both British and German salvage teams in wartime to simply fill in the hole where a fighter had crashed without bothering to remove the body of the pilot has been demonstrated. The reason for this is only partly wartime callousness, which results in the dead men being misleadingly posted as 'missing'. The main reason is that mechanical excavators, which can dig up a deeply-buried wreck in half an hour if necessary, were not available in wartime. The job would have had to be done with picks and spades, and could have taken days. This is probably the true explanation of why the Widley Hurricane and Willius' Focke-Wulf were in effect 'written off' without ceremony. But is it really so wrong, long after, to give the dead men names?

Chapter 14

THE LOST AIRLINERS

The two Tudors and the two Comets: 1948–1954

'No more baffling problem has ever been presented for investigation,' concluded the official Accident Report. 'The Court has not been able to do more than suggest possibilities, none of which reaches the level even of probability.' *Star Tiger*, an Avro Tudor IV of the recently formed British South American Airways Corporation, had vanished on 30 January 1948, when some 340 miles out from Bermuda after a flight of nearly 2000 miles from the Azores. The airliner had been within a few hours of her destination. Her last message had acknowledged receipt of a bearing taken by Bermuda. There had not been the least hint of anything wrong. But she had vanished, and without trace. A thorough air search by twenty-five aircraft had at once been made along her track and had been continued for five days. But not a trace of the missing aircraft or of the thirty-one men, women and children aboard had been discovered.

Both the aircraft and the airlines concerned represented an early attempt to bridge the awkward gap between war and peacetime aviation. The Tudor had been conceived in 1943 as a civilian version of the successful four-engined Lancaster bomber, but British Overseas Airways Corporation had turned it down eventually because of delays in delivery plus a realisation that it might prove uneconomic. As their version carried only twelve passengers (admittedly, they could all lie down and sleep the long flight hours away), this was hardly surprising.

At the end of the war a new airline was formed privately by a group of demobbed airmen drawn from the elite of Bomber Command, the Pathfinders, with the legendary Don Bennett as their chief executive. Called British South American Airways, the new corporation would pioneer the long and lonely routes over the South Atlantic, a challenge even to the heavily-decorated men of the crack navigational and target-marking force of the RAF. They believed that the Tudor, based on the aircraft they were used to flying, would be suitable for economic airline operation provided that the flight engineer's position was dispensed with and accommodation provided for thirty-two sitting passengers. The cabin was rearranged, the fuselage lengthened by six feet, and the result was the Tudor IV. The first three machines were delivered late in 1947 and named *Star Panther*, *Star Lion*, and *Star Tiger*. When the first disappearance occurred it involved the newest machine, *Star Tiger*, built less than three months earlier.

The route was from London to Havana in Cuba at an average speed in still air of 165–175 nautical miles per hour. As the Tudor had a flight endurance of sixteen hours, refuelling and resting stops were necessary. The first stage was to Lisbon, where an overnight stay was made. Leaving Portugal next morning, the airliner would fly to the Azores, where it would refuel again. After one and a quarter hours on the ground, it would take off on the most critical stage of the whole route, the 1961 nautical miles of entirely empty ocean between the Azores and Bermuda. There were no land masses, no islands, not even a single weather ship. And the navigational target, the Bermudan group of islands, covers an area of only twenty square miles. Cape Hatteras, the nearest part of the North American continent, is 580 miles to the northwest.

Historically, Bermuda and the Azores were for centuries on the homebound route of the Spanish treasure fleets, because the prevailing winds in that area of ocean were from the southwest; these winds blew them home to Portugal and Spain. Necessarily, any aircraft outward-bound from Europe on that route must most often meet headwinds, sometimes prohibitively strong winds. The old Spanish outward-bound route for sailing ships lay far to the south, to take advantage of winds which there blew them southwest. The airline did not have the option of taking that route; for half the time their machines were forced to fly against the weather pattern. Superficially, they had the range – in still air, an endurance of sixteen hours covering a distance of about 2900 nautical miles gave them nearly a thousand nautical miles in hand. But that melted away, when one deducted two hours reserve in case of difficulty in finding or getting into the airport, and then slapped on perhaps a 50 mph headwind for nearly 2000 miles. Consequently, in each flight plan the pilot had to indicate a 'point of no return', after which it was too late to turn back or divert to an airfield in North America; after that point, he was committed to going on to Bermuda whatever the weather conditions might be.

Star Tiger left London airport for Lisbon on 27 January 1948, with twenty-three passengers and a crew of six. She had logged 576 hours flying time since delivery and had just undergone a complete overhaul. Nevertheless, she was to develop ineradicable defects on this flight – failure of the cabin heating and trouble with one of the compasses.

The Tudor carried three pilots to fly and navigate the aircraft, which theoretically allowed for one man to be off duty at any one time. There was no engineer and there was only one radio operator, who might therefore be on duty continuously for up to twelve hours; and with a short rest, for much more than that. There were also two 'Star Girls' – stewardesses.

All the crew were very experienced. Captain B.W. McMillan, the commander, had more than 4500 hours in his logbook, 1673 of them in command of the Corporation's aircraft, the rest with the RAF in which he had attained the responsibility of 'master bomber', the kingpin of the Pathfinders who actually directed the raid from above the flaming target and kept the aiming point continually marked up. Sometimes it was a short life, but McMillan, a New Zealander, had survived to make a successful transition from war to peace. The First Officer, David Colby, had logged more than 3000 hours; he had been a Pathfinder too. The Second Officer, Mr C. Ellison, had logged nearly 2000 hours. The oldest man in the crew was 'Tucky', the infectiously humorous radio officer, Robert Tuck. He had served fourteen years as a radio officer at sea and for most of the war had been senior radio officer of the North Atlantic Control at Prestwick in Scotland. This particular route called for exactly that experience of astronomical navigation which Tuck possessed.

The Tudor crossed the Bay of Biscay at 21,000 feet with the cabin heating out of order and the windows frozen up from the inside; conditions of bitter discomfort which had to be endured hour after slow hour. Repairs were made during the overnight stop in Lisbon, but the heating failed again during the next stage, the flight to the Azores. The faulty compass had gone out of order again also, and take-off had been delayed by the failure of a priming pump for the port inner engine. Some of the passengers wrote bitter little letters to friends; two more people had boarded at Lisbon, bringing the total to twenty-five.

After the long flight from Lisbon, the Tudor should have refuelled and then taken off again for the even longer flight to Bermuda, regardless of crew fatigue. But a 60-knot wind was screaming across the airfield on Santa Maria in the Azores when Captain McMillan landed. Such a headwind put Bermuda out of reach. Nevertheless, the decision to stay overnight at the Azores as well as at Lisbon was not made by McMillan on his own but in consultation with another captain of the airline, Frank Griffin, whose task was planned to fit in with his own. Griffin commanded a four-engined freighter which, developed more directly from the Lancaster bomber, was called a Lancastrian. Some of her cargo was to be transhipped at Bermuda to the Tudor for the short last leg to Havana. The endurance of the Lancastrian was nineteen hours as against the slightly faster Tudor's sixteen hours, and Griffin probably could have made Bermuda in spite of the headwind; but there was no point in his arriving ahead of the Tudor and good safety reasons why the two aircraft should fly in company.

Next morning, 29 January, the wind had lessened and after studying the weather forecasts the two captains decided to go. At height, the winds were stronger, so both would fly at 2000 feet (and perhaps be unable because of cloud to get a star-fix); and to allow for handling time on the ground at Bermuda, there would be an hour's interval between the two aircraft. The Lancastrian with its greater endurance and safety margin would fly in the lead, sampling the actual weather (as opposed to forecast weather) and passing reports back to the Tudor an hour behind. At 2000 feet, the failure of the cabin heater would not be noticed. It was January, but they were in the warm latitudes.

Captain Colby in the Lancastrian took off at 2.22 pm, followed by the Tudor at 3.34 pm. Contrary to regulations, Captain McMillan took off with the Tudor overweight by some 1000 lbs. Theoretically, he had the choice of either dumping some passengers or taking off with tanks less than full. He was reluctant to dump the passengers, and said he would accept less fuel, but in the event he instructed the ground engineer to 'fill her up to the gills'. Clearly, his experienced appreciation was that while to take off overweight involved a slight risk, attempting Bermuda with tanks less than full would be foolhardy. The overweight would soon disappear in flight as the petrol was consumed.

On the other hand, on this route westerly winds much stronger than those forecast could be encountered. Twice *Star Tiger* had been forced to divert to Newfoundland while attempting Bermuda, and another Tudor had actually landed at Bermuda with the fuel gauges showing 'nil' and the air-sea rescue organisation fully alerted for a probable ditching.

Take-off had been made on the basis of an Azores forecast and both aircraft initially worked Santa Maria radio until eventually they were able to contact Bermuda ahead. Over a stretch of some 2000 miles, the weather was unlikely to be uniform; nor was it. The trouble was that it began to differ increasingly from the forecast, and for the worse, at about the times that critical decisions had to be taken. Most of the radio work was done by the Lancastrian, acting as 'pilot' for the Tudor following 200 miles and one hour behind. While the Tudor sent out mainly position reports, it was Griffin's aircraft which continually asked for wind speeds and directions in the various weather zones ahead. Bermuda, although nominally British, had become an American base as part of the wartime Lend-Lease deal between Roosevelt and Churchill; and as the Accident report noted, there was a 'striking contrast between the delays involved in obtaining a reply through the ordinary channels and the speed with which the US Army Air Force answered'. When the Lancastrian asked for a forecast at 5.52 pm, the Americans at Kindley Field were sending the answer at 6.06 pm. This was: In zone 12, winds at 2000 feet were predicted for 22 knots; in all zones except 13 (round Bermuda itself) all winds were stronger than had been forecast by the Azores.

This agreed with the calculations of the Lancastrian's navigator, made as the aircraft was passing through zones 7 and 8. Instead of the 30- to 40-knot winds foretold at the Azores, actual wind speed was 55 knots. That put back her estimated time of arrival at Bermuda by one hour, to 13.28 next day, a flight time of 13 hours 28 minutes. At 6.49 pm the Lancastrian passed this news to *Star Tiger*, as her flight time was bound to increase also. And at 7.45 pm, the Lancastrian passed on the details of Kindley Field's forecast to Roger Tuck in the Tudor. Captain McMillan would know that his margin of safety had been cut by one hour.

At 1.26 am (GMT) on 30 January the navigator of the Lancastrian obtained an astral fix. The reason for flying this stage largely in darkness was that star-fixes, if obtainable, were more reliable than dead reckoning and observations of drift. The aircraft was sixty-eight miles north of where it should have been, on a course to miss Bermuda completely. The reason, according to her navigator's calculations, was that instead of the 'light and variable' winds forecast by the Azores, a south-westerly wind of 48 knots was pushing the Lancastrian bodily north of its intended track. Her captain began to ask Bermuda for local forecasts from 2 am onwards, but it was not until 2.42 that he got a forecast of 'SSW 20 knots'.

Meanwhile *Star Tiger* had steadily been transmitting position reports every hour on the hour, had put back her estimated time of arrival to 5 pm because of increased headwinds, and by her 2 am position report had shown that her navigator also had obtained an astral fix which had revealed a drift off course to the north. But as her position at that moment was known with certainty, a new and more accurate course would have been set.

The next hourly report from the Tudor, of her position at 3 am on 30 January, was promptly passed by Tuck to Bermuda at 3.02 am. Two minutes later he requested the Bermuda station VRT to take a radio bearing on the Tudor, as the 3 am position had been made by dead reckoning alone. He depressed the transmitting key to give a continuous signal for VRT's directional aerial to work on, but the Bermuda station was unable to obtain a satisfactory reading. At 3.15 am Tuck tried again and this time Bermuda obtained a first-class bearing of 72°. 'First-class' meant with an error of no more than 2°. Even if the wind fluctuated greatly in strength and direction, and overcast skies prevented another astral fix being taken, a series of such bearings would bring *Star Tiger* safely to Bermuda with a fair margin of fuel left.

Star Tiger acknowledged receipt of Bermuda's message regarding the class one bearing, and then fell silent. For ever.

Strictly speaking, the ground operators were required to declare an emergency after thirty minutes loss of contact with an aircraft. The VRT operator called *Star Tudor* again at 3.50 am, thirty-five minutes after the last acknowledgement, and got no reply. Assuming she had switched over to direct voice communication for the last lap, he asked Bermuda Approach Control if they had *Star Tiger*. They had not, so he called her again at 4.05 am. No reply. He called once more at 4.40 am, twenty minutes before she was due to land, and still there was no reply. Then he declared a state of emergency.

At 4.55 am the Search and Rescue Section of the USAAF at Kindley Field were alerted and by 7.16 am (which was 3.16 am by Bermuda time) an American Fortress equipped with a radar scanner was airborne. That day, twenty-five other aircraft joined her, including the British Lancastrian which had arrived safely. The search went on until nightfall on 3 February. No sign of the Tudor had been found, either by aircraft or by surface vessels. Even more surprising was that no distress call had been heard. Many operators were listening out on her frequency and would have heard had Tuck even begun to transmit. *Star Tiger* had vanished not merely without trace but without a sound.

The evidence put before the Court only deepened the mystery. It ruled out completely the possibilities of atmospheric disturbance and electrical storms, because other aircraft in that area at the time reported stable weather conditions. In particular, the Lancastrian, trail-blazing one hour ahead, had found that cloud base was above 2000 feet and visibility good enough for Captain Griffin to see the lights of Bermuda when he was twenty-five miles out. Of the remaining possibilities, ran the Report, 'none reaches the level even of probability', and it concluded: 'it may truly be said that no more baffling problem has ever been presented for investigation.'

Twelve months later another Avro Tiger, *Star Ariel*, disappeared without trace near Bermuda.

On this occasion the Court wasted no time in trying to find out what had happened; in trying to explain the inexplicable without a scrap of evidence to go on. The bulk of the Report was devoted to the confused radio control procedures in the area, which had resulted in a delay in this case also in initiating a search: this was a matter which might be remedied and save life in the future, whereas the seven crew and thirteen passengers of *Star Ariel* were gone beyond recall.

Star Ariel was an Avro Tudor IVB, with more powerful Rolls Royce Merlin engines and an engineer's panel. Bar one man, the crew were all very experienced. Captain J.C. McPhee, the commander, had served with the Royal New Zealand Air Force and BOAC previously, and his total flying hours exceeded 3000. His two co-pilots were similarly well qualified, as was the Engineer Officer. The Radio Officer had exceptional skills, including above-average ability to repair his equipment. The 'Star Girl', Miss J.B.

Moxon, who had been a VAD nurse, had logged 868 hours in the air. The solitary novice was the Steward, who had flown only seventy-two hours; but he had for years been an RAF messing sergeant.

The aircraft left London on 13 January 1949 for Jamaica, via stops in Newfoundland, Bermuda and the Bahamas. It was back in Bermuda on the homeward flight on 17 January, when a change of plan sent it back to Jamaica. This was because an incoming Tudor on the scheduled westbound service had been grounded with engine trouble. *Star Ariel* was to replace her.

Captain McPhee took off from Bermuda at 12.42 hours with petrol for ten hours' flight aboard; his estimated time of arrival at Kingston, Jamaica, was 18.10 hours, so he had an ample reserve. At the height he intended to fly, 18,000 feet, visibility above cloud was unlimited. The forecast wind was northerly 36 knots, the actual winds close to that. There was no question of icing, and the chance of turbulence was almost nil; in fact, 'there were no weather complications at all', concluded the Accident Report. Indeed, on this occasion the wind was actually helpful, drifting the aircraft southwards towards the Caribbean.

At 13.35 hours, some three quarters of an hour after take-off, *Star Ariel* sent off a message to three addresses – La Guardia Field, New York, and her own airline offices in both Bermuda and Jamaica. It read: 'I departed from Kindley Field at 12.41 hours. My ETA at ZQ JK 18.10 hours. I am flying over 150 miles south of Kindley Field at 13.32 hours. My ETA at 30° N is 13.37 hours. Will you accept control?'

In brief, the aircraft had left the Bermuda area of control and was asking Jamaica to take over. Seven minutes later, at 13.42, *Star Ariel* again sent a message to multiple addressees: 'I was over 30° N at 13.37 hours. I am changing frequency to MRX (freq. 6523 kc/s).' This message from the aircraft was acknowledged and the acknowledgment was authority for Captain McPhee to change radio frequency to Kingston, Jamaica, his destination.

That was the last message received by any known radio station. *Star Ariel* never spoke again.

Because she vanished between control stations there was some confusion on the ground as to who was responsible for the aircraft, and search procedures were not initiated until after 19.00 hours. Only two aircraft took off that evening. But from 18 January to 23

January, between seventy and eighty aircraft were airborne per day carrying out a highly organised search along the route *Star Ariel* would have taken, and to sixty miles on either side of it. New York Coastguard coordinated the search, which involved American rescue squadrons from Florida and Puerto Rico. The results were the same as for *Star Tiger* – another Avro Tudor had vanished without trace near Bermuda.

Lord Brabazon of Tara, a pioneer airman and aviation expert, undertook an investigation into the Tudor IV design from the safety viewpoint. He gathered a body of experts who carried out a programme of research and experiment. They studied the records of the type and restudied the evidence regarding *Star Tiger* as well as the new evidence (or lack of it) for *Star Ariel*. This committee could find no probable causes for the two disasters. The reason, whatever it was, had to be fairly radical and virtually instantaneous, so that not even a broken SOS could be sent. Lightning was suggested, as was static electricity, and the presence of noxious gases such as carbon monoxide. But without wreckage there was no evidence.

Later experience made some people believe that the probable cause may have been a complete electrical failure in the aircraft. This would silence the radio and put many of the vital flight instruments out of action. In the already critical circumstances of *Star Tiger*'s flight across empty ocean to Bermuda, this might well explain her disappearance. But hardly that of *Star Ariel*, flying in good weather and excellent visibility towards an almost continuous chain of inhabited islands, and moreover with ample fuel reserves.

But in due course both these disappearances went to help build the legend of the 'Bermuda Triangle', an area of supposed magnetic and sometimes frankly magical anomalies between the Sargasso Sea and Florida in which both ships and aircraft vanish without any wreckage or oil slicks being found. However, when the last messages from the lost Tudors are mentioned to 'prove' the Triangle theory, they are often badly misquoted or bear no resemblance to the actual messages sent. The story seems to have been suggested by another truly mysterious disappearance a few years before, on 5 December 1945. A complete flight of five TBM Avenger torpedo bombers became lost over the sea, were unable to find their way back to land, and vanished without trace – all five of them. This could well have been the result of cumulative navigational errors by the leading crew which necessarily took all the others to their doom. However, to compound this mystery, a Martin Mariner flying boat sent out to search for the lost Avengers itself disappeared, also without trace.

When an aircraft is never found, mystery there must always be. Sometimes there is such pressure to establish the reason for an accident that it leads to the discovery of the missing aircraft. This was the case with the historic DH Comet G-ALYP (usually known as *Yoke Peter*), which on 2 May 1952 had inaugurated the world's first service by a pure-jet airliner in a brave bid by Britain to capture world markets from America by making a vast technological leap forward. Prewar British machines could not compete, partly from outdated designs, but largely because there was in North America a large home market for long-range airliners. In the immediate postwar period the Americans were supreme. Then the introduction of the Comet put Britain ahead by more than six years. All this was at stake when on the morning of Sunday, 10 January 1954, *Yoke Peter* took off from Rome for London. At 10.50 she reported being on track between the Orbetello and Elba radio beacons, and climbing to 36,000 feet. A few minutes later her radio operator was talking to another British Overseas Airways machine, an Argonaut, when his voice was wiped out in mid-sentence.

Giovanni di Marco was out fishing south of the Isle of Elba in the Tyrrhenian Sea when he heard the whine of a high-flying aeroplane invisible above cloud. Then three explosions occurred, rapidly, in succession. There was a moment of silence, until he saw a silver object flash down out of the clouds, smoking, several miles distant. It struck the sea and a great white cloud of water foamed up. Marco drove his boat towards the scene, but all he and the other Italian fishermen found were fifteen bodies drifting lifelessly with the current. There were no survivors from the thirty-six passengers and crew of six, and only a few small fragments of the Comet left on the surface.

Lieutenant-Colonel Lombardi, the harbour master at Elba, on his own initiative located sixteen witnesses of the crash and with their evidence was able to plot sixteen lines of observation going out to the south, most of which intersected fairly closely; only one witness was way out to the west. An aircraft belonging to Skyways

Ltd took a photograph of the fishing boats grouped around the drifting wreckage with part of Elba itself in the picture. This information fixed the general area of the accident, with a bias to the south due to the current moving the bodies. The sea in this area was between 450 and 600 feet deep.

Compared to the disappearance of *Star Tiger* and *Star Ariel*, this was precision indeed; but to actually locate the wreckage in what was only a general area and to have recovered it with the primitive means available in the 1950s would be to ask a good deal. Yet it was necessary, indeed critical for British civil airline development. Clearly, this accident had nothing to do with bad weather. While the radio operator was actually talking, the Comet had exploded under him. Why?

Assuming that asdic (as sonar was then called) could pick up all likely wreckage in the general area, could divers go down to confirm that what had been found was indeed the Comet? When the question was put to Victor Campbell, the Senior Admiralty Salvage Officer at Malta, he replied: 'Provided it is found in 180 feet of water, we'll recover it for them. 200 feet at the outside. Neither our divers nor compressors will get down farther than that.'

While he was giving this answer a Cabinet meeting was being held in London to debate what should be done. Mr Lennox-Boyd was delegated to telephone the Commander-in-Chief, Mediterranean, for advice. This officer, Lord Mountbatten of Burma, was in his bath at his home near Romsey in Hampshire. Dripping wet, he took the call and heard Lennox-Boyd say that it was vital to salvage the Comet and did he know of a salvage firm capable of doing it? After Lord Mountbatten had checked the depth with him, the politician having confused fathoms with feet, a sixfold difference, Lennox-Boyd said: 'Sorry. I meant 600 feet. What salvage firm do you think I could approach?'

'My own, of course,' replied Mountbatten.

And so the Mediterranean Fleet got the job.

Lord Mountbatten's chief-of-staff, Commodore Woods, passed on the bad news to his two top salvage experts, Commander Gerald Forsberg and Victor Campbell. 'As the C.-in-C. so succinctly put it, we have to divide the job into two parts. First, to locate. The aircraft is somewhere in a hundred square miles south of Elba. Second, to salve. It is somewhere in about 400 or 500 feet of water, which is more than twice the depth our present gear can tackle. Easy, isn't it?'

Anti-submarine frigates of the Mediterranean Fleet probed the area with asdic and located more than a dozen contacts, which were buoyed for underwater investigation. Two slow and ungainly ships were employed on this work: HMS *Barhill*, which laid heavy moorings all round each contact, and the fleet auxiliary *Sea Salvor*, which utilised them to move herself back and forth across the area on six mooring cables controlled by six winches. A seventh winch let down a rigid observation chamber with a diver inside or, alternatively, an underwater television camera. An eighth winch lowered a weight in front of the observer (or TV camera, if used instead) as a directional guide. Two winches were required for the grab which would recover parts of the wreckage, plus another for the derrick topping-lift. Eleven winches in all, so that the decks were a potentially lethal cat's cradle of heavy cables under tension; and as there were exacly eleven deckhands in the Sea Salvor's crew, there was one man per winch and everyone was on duty all the time.

But the work was also exciting. Both the TV camera and the observation chamber were new, loaned by civilian firms in England, and with the former particularly one could regard the work as exploration. 'It is uncanny to think you are looking somewhere where no human eye has looked before,' wrote Gerald Forsberg.

I have myself dived to 165 feet in the Tyrrhenian Sea, but in summer when it is like swimming in champagne. This January, the continual gales stirred up the bottom so that maximum visibility was only twenty-five feet and often it was twelve feet or less. Assessing contacts took much longer in those conditions, quite apart from the fact that the TV camera, unlike a diver, cannot look round corners. After being kept in port for days by a blow, the salvage vessels would come out to moor above the contact to find after days of work that it was a wartime mine, a sunken dan buoy, a recent wreck or, once, a very ancient wreck indeed, a Greek or Roman freighter carrying a cargo of amphorae. For a moment, the face of a sunken statue appeared on the TV screen, to be glimpsed by human eyes for the first time in perhaps two or three thousand years. But they let it go by in the haste to find *Yoke Peter*.

Of course, no one was actually looking for a complete aircraft, and indeed what was found of the Comet at first made the seabed look 'as though someone had up-ended a waste-paper basket'. But there were larger pieces, although the initial recovery was made in the simplest possible manner. A trawler was used to lay a warp in a circle and bring the ends back to *Sea Salvor*. The device lacked nets and otter-boards and was only lightly weighted: it should not have succeeded. But the chain caught round a chair, the chair snagged in some electric cables, and the cables were still attached to the after pressure-dome of the Comet. When the whole fantastic collection was swung inboard, it was found that the navy had rescued the ladies' and gentlemen's toilets from 400 feet down. This 'lode' proved to be entirely from the aft section of the airliner, and then it petered out.

'After the Lord Mayor's Show comes the muck wagon,' was a prediction much favoured by the salvors, because it worked the other way round, too. After great disappointments and several gales, a new contact proved to be the centre-section of the Comet, some sixty-five feet long. Nearby were the four jet engines, in which the technicians were greatly interested in case the fault lay with them.

Finally, in a new position suspiciously close to a wartime minefield, a contact was snagged by the trawlers at the end of March. The day was calm, the visibility was twenty-five feet, but the diver sent down in the observation chamber could not at first make out what the contact was, because it was enfolded in layers of thick trawl net. By venturing close in to the tangle of nets and warps, the diver decided it was part of an aircraft. As indeed it was. The grab raised to the surface the biggest single piece of Comet so far, the entire forward end, from the nose right back to the wings. The vital clue to the disaster proved to be hidden among this part of the wreckage.

Sea Salvor had finished her job and was due to return to Malta on 9 April, leaving the trawlers to pick up the smaller pieces of wreckage still on the bottom, when a message was handed to Commander Forsberg: 'From Commander-in-Chief, Med. Proceed to search for BOAC Comet G.ALYY missing on flight from Rome to Cairo.'

Yet another of the new jetliners had gone down into the sea, this time off Naples to the south of Elba. All Comets were grounded until the wreckage raised from the first crash had been put together and examined. The experts already had a theory. If sabotage was excluded, then the obvious cause of an explosion would be some small fracture occurring in the pressurised cabin due probably to metal fatigue. The air is so thin at 30,000 feet that it has a pressure of only 4 lb per square inch, compared to the 14.7 lb per square inch at sea-level. The pressure in the cabin of an airliner is artificially maintained at about 11 lb per square inch. If even a small crack occurred at some part of the cabin, the air inside the machine would rush out through it, completely rupturing the aircraft's skin in an 'explosive decompression'.

That was the theory. But it could only become fact if the salved wreckage of the Comet *Yoke Peter* included the fragment which had fractured. And it did. The experts found traces of metal fatigue and subsequent rupture near a window at the forward end of the fuselage where the direction-finding aerial was housed.

Once this was established without doubt, the weakness could be remedied and the Comets put back into service. The delay and the lack of confidence engendered by the two accidents seriously affected British aviation, but in the 1980s Comets are still flying over the Mediterranean.

Could something like this have happened to the two Avro Tudors flying in the Bermuda Triangle? The answer, almost certainly, is no. To start with, they were new aircraft. *Star Tiger* had flown only 575 hours, *Star Ariel* 340 hours, whereas *Yoke Peter* had logged 3681 hours, far more than any other Comet in the B.O.A.C. fleet at the time. Then *Star Tiger* was at 2000 feet, where pressurisation is not necessary. The jets, flying close to the speed of sound at heights well over 30,000 feet, were a whole new world away from the old propeller-driven aircraft in the hazards they were to encounter.

Where the two Avro Tudors crashed, and why, will now never be known. Unless some freak discovery of the future, a million-to-one chance, should turn up the answer.

SOURCES

Chapter 1:

LOST IN MYTH AND LEGEND

The Palace of Minos at Knossos by Sir Arthur Evans, Macmillan, London, 1921–35.

The Bull of Minos by Leonard Cottrell, Pan, London, 1955.

The Minoan World by Arthur Cotterell, Book Club Associates, London, 1979.

Knossos by Chr. Mathioulakis & N. Gouloussis, Athens, 1978.

Island of Scopelos: Historical & Archaeological Study by D. Sampson, Athens, 1968.

Chapter 2:

THE ICE FLIGHT OF THE "EAGLE"

Med Ornen mot Polen, Bonniers, Stockholm, 1930.

The Andrée Diaries, Bodley Head, London, 1931.

Unsolved Mysteries of the Arctic by Wilhjalmur Stefanson, Harrap, London, 1939.

De Döda Pa Vitön by E.A. Tryde, Bonniers, Stockholm, 1952.

Chapter 3:

THE MONTROSE GHOST

'On the Montrose Ghost' by C.G. Grey, *The Aeroplane*, 29 December 1920.

'The Montrose Ghost' by Anon, *The Aeroplane*, 16 October 1942.

'The Montrose Ghost' by Sir Peter Masefield, *Flight*, 21 December 1972.

'Montrose Ghosts' by G.M. Macintosh, *Flight*, 4 January 1973.

Wings Over Westminster by Lord Balfour, Hutchinson, London, 1973.

Personal communications: Lord Balfour, Sir Peter Masefield,

G.M. Macintosh, Royal Aircraft Establishment Library, Farnborough, Michael Jerram

Chapter 4:
WHO KILLED THE ACES?
Max Immelmann: Eagle of Lille by Franz Immelman, Hamilton, London, 1940.
Knight of Germany: Oswald Boelcke by Johannes Werner, Hamilton, London, 1933.
Flying Dutchman by Anthony Fokker, Routledge, London, 1932.
Captain Albert Ball by R.H. Kiernan, Hamilton, London, 1933.
Sagittarius Rising by Cecil Lewis, Davies, 1938.
'The Red Air Fighter' by Manfred von Richthofen, *The Aeroplane*, 1918.
An Air Fighter's Scrapbook by Ira Jones, Nicholson & Watson, London, 1938.
Aces High by Alan Clark, Fontana/Collins, London 1974.
Ace of the Iron Cross by Ernst Udet, Doubleday, New York, 1970.
Fighting Planes and Aces by W.E. Johns, Hamilton, London, 193?.
Who Killed the Red Baron by P.J. Carisella and James W. Ryan, Avon, New York, 1979.
'Sky Fighters of World War I' (*Cavalier Magazine*, Fawcett Books, 1961).
The Aces by Frederick Oughton, Spearman, London, 1961.
'Georges Guynemer' by Douglas Whetton, *War Monthly* No. 48.
The Anatomy of Courage by Lord Moran, Constable, London, 1945.
'The Riddle of Richtofen's Death' by Arch Whitehouse, *Flying Aces*, January 1935.
'Who Slew Baron von Richthofen?' by Arch Whitehouse, *Flying Aces*, December 1935.
Correspondence re Capt. Ball, *Popular Flying*, December 1934.
Correspondence re Capt. Ball, *Daily Telegraph*, May–June 1962.

Chapter 5:
THE HIGH SPEED FLYER
'Schneider Trophy' by W/Cdr. A.H. Orlebar (Seeley Service).
Unpublished Manuscript by A/Cmdre D. D'Arcy A. Greig.
Taped Lecture by L.S. Snaith (1931 team), courtesy W/Cdr. D.E. Bennett.
Reports: *Southern Daily Echo*, *The Aeroplane*, *Flight*.

Interviews and/or personal communications: A/Cmdre D'Arcy Grieg, W/Cdr. D.E. Bennett, A.C. Munnings, R. Sevier, L. Bartholomew, D. Pragnell, W.C. Etherington, T.L. Banks, E.W. Gardener, C.A. Mann, M. Young, Mrs. L.M. Watts, T.R. Hiett, Miss A.M. Kinkead.
Assistance with documents: Royal Aircraft Establishment Library, Farnborough; Central Library, Southampton; Central Library, Portsmouth.
Personal Diary: Alexander McKee

Chapter 6:
VANISHED

My Flying Life by Sir Charles Kingsford-Smith, Aviation Book Club, London, 1939.
Pacific Flight by P.G. Taylor, Hamilton, London, 1940.
'Some Technical Aspects of Sir Charles Kingsford-Smith's Flights' by Sir Lawrence Wackett, 16th Memorial Lecture, Sydney, September 1974.
The Search for Amelia Earhart by Fred Goerner, Bodley Head, London, 1966.
World Flight: The Earhart Trail by Ann Holtgren Pellegreno, Iowa State University Press, 1971.
Great Mysteries of the Air by Ralph Barker, Chatto & Windus, London, 1966.
Amy Johnson by Constance Babington-Smith, Collins, London, 1967.
Scott's Book by C.W.A. Scott, Hodder & Stoughton, London, 1934.
The Coal-Scuttle Brigade by Alexander McKee, Souvenir Press, London, 1957.
Ice Crash by Alexander McKee, Souvenir Press, London, 1979.
Personal communications: E.P. Wixted, Queensland Museums; Ann Welch; RAF Museum, Hendon; Dr. W.T.K. Cody; Stuart V. Tucker; Michael Jerram.

Chapter 7:
THE MAN WHO WAS NEVER MISSING

Helmut Wick: Das Leben eines Fliegerhelden by Joseph Grabler, Verlag Scherl, Berlin, 1941.

Major Wick: Das Vorbild des Deutschen Jagdfliegers by Walter Zuerl, Verlag Zuerl, München, 1941.
Enemy in the Sky: My 1940 Diary by Air Vice Marshal Sandy Johnstone, Kimber, London 1976.
Combat Reports, Operations Record Books 152, 213 and 609 Squadrons (PRO, Kew).
Personal Communication: Hugh Dundas.
Personal Diary: Alexander McKee.

Chapter 8:
THE GHOSTS OF WIGHT
Interviews or personal communications: Violet, Hubert & Garry Brown, John Cleaver, Mrs. Winifrid Drudge, Miss Vera M. Thorne, Mrs Amy R. Saunders, David A. Saunders, T.C. Williams, Rex Burton, Oscar H. Smith, R.J. Atkins, F.T. Aylett, Brian Warne, Stewart Pont, C. Michell, F.R. Rolf, Reg. C. Eve, IOW County Archivist, PRO, Kew.
Personal Diary: Alexander McKee.

Chapter 9:
NO CHRISTMAS IN CONWAY STREET
Interviews or personal communications: Frederick Gerald Bishop, Frederick James Kistle, William Wills, L.R. Stubbington, P/C Kerby, Ronald Walker, Mrs Eileen Skeates, R.J. Cuss, Walter Jeakes, C.M. Jeram, R.P. Houghton, Frederick James Pepper, Alan David Dart, Mrs Elsie Pitman, Mrs Edith Sawyer, Lionel Whitelock, Mrs Lilian Owen, Wally Farnes, George William Batterham, Donald McGregor, George Rex, Cecil Thompson, Cecil Boniface.
Portsmouth City Records Office: Civil Defence Wardens Service War Diary; Official Municipal History of the War.
Public Record Office, Kew: Home Security West Region Daily Appreciation; Home Security Intelligence Summary; Admiralty – Enemy Air Attacks, Portsmouth.
Bundesarchiv, Militärchiv, Freiburg: Reports, Luftflotte 3.
Portsmouth *Evening News*: 1965, 1972, 1979; booklet *Smitten City*.
Personal Diary: Alexander McKee.

Chapter 10:
DEATH OF A PRIME MINISTER
Accident by David Irving, Kimber, London, 1967.

The Assassination of Winston Churchill by Carlos Thompson, Smythe, Cross, 1969.
'The Death of General Sikorski', *After the Battle No. 20.*

Chapter 11
ABANDONED

Recollections of an Airman by Lt.-Col. L.A. Strange, Aviation Book Club, London, 1940.
Lighter than Air by Lee Payne, Barnes, N.J., Joseloff, London, 1977.
Lighter than Air Flight by Jack Pearl, ed. Lt.-Col. Glines, Watts, NYC, 1965.
The Archives of: Christopher Elliott, Michael F. Jerram.
Personal Diary: Alexander McKee.

Chapter 12:
INTO THIN AIR

The Lost Prince: Young Joe, the Forgotten Kennedy by Hank Searls, New American Library, New York, 1969.
Aphrodite: Desperate Mission by Jack Olsen, Putnam, New York 1970.
'Death of a Kennedy' by Samuel Eliot Morison, *Look* Magazine, February 1962.
Kennedy's Fatal Mission by F/Lt Tony Fairbairn and Stewart Evans *Aircraft Illustrated*, March 1977.
The Archives of Christopher Elliott and Stewart Evans.
Personal Communications: Mick Muttitt, W.S. Jarvis, Cdr. John F. Burger Jr USN, Robert Ball, W.J. Wallis.
Archives of HMS *Centurion*
The Dam Busters by Paul Brickhill, Evans, London, 1951.
V 2 by Maj-Gen Walter Dornberger, Hurst & Blackett, 1954.
Crossbow and Overcast by James McGovern, Hutchinson, London, 1965.
Air Spy by Constance Babington-Smith, Ballantine, New York, 1957.
The Defence of the United Kingdom by Basil Collier, HMSO, London, 1957.
The Canadian Army 1939–1945 by Col. C.P. Stacey, Ottawa, 1948.
'The V-Weapons' *After the Battle No. 6*, 1974.
Personal Diary: Alexander McKee.

Chapter 13:
BACK FROM THE BED OF THE SEA
Archives & Personal Communications: Lt Col A.P. De Jong, RNAF, Luchtmacht-Voorlichtingdienst, The Hague; G.J. Zwanenburg, RNAF, Koninklijke Luchtmacht Salvage Team, Soesterberg.
Geschichte Eines Jagdeschwaders, JG.26 by Joseph Priller.
War Over England by A/Cmdre L.E.O. Charlton, Longmans, Green, London 1938.
Holland and the Delta Plan by J.S. Lingsma, N.V. Uitgeverij Nijgh & Van Ditmar, Rotterdam, 1963.
'Zuyder Zee Archaeology' by G.D. van der Heide, *Antiquity & Survival*, The Hague.

Chapter 14:
THE LOST AIRLINERS
Ministry of Civil Aviation Reports: 1948 and 1949.
Great Mysteries of the Air by Ralph Barker, Chatto & Windus, London, 1966.
Up She Rises by Frank W. Lipscomb & John Davies, Hutchinson, London, 1966.
'Operation Elba Isle' by Gerald Forsberg, *Blackwood's Magazine*, September 1954.
Personal Communications: Sir Peter Masefield, Ralph Barker.